Johannes Wild
Lohmaierstr. 7
94405 Landau
Germany

3dtechworkshop@gmail.com

Caution: This book is for educational purposes only. Electricity, especially alternating current and high voltage current is dangerous to life. Seek professional advice before carrying out any practical work.

Foreword

Thank you so much for choosing this book!

A warm welcome! Are you looking for a simple and understandable introduction to the basics of electrical engineering and electronics?

Then you are well advised with this book! I am an engineer (M.Eng.) and would like to teach you the basics of electrical engineering and electronics. This book offers you a well understandable, intuitively structured and practical introduction to the world of electrical engineering!

What is current and what is voltage? What is charge? What is power? How does an electric motor work, what is the difference between direct current and alternating current? This handbook of electrical engineering not only answers these questions, but also covers many other topics in depth and detail. In addition to important basic terms and principles, you will also learn, for example, how to analyze electrical engineering circuits, what a transistor (bipolar and MOSFET) is, and how an RLC circuit is designed. We will also look at what happens when you place a coil in a magnetic field and what practical applications these basic principles have in our modern world.

This fundamentals book is aimed specifically at all those who have no or only previous school knowledge in electrical engineering and electronics or already have knowledge and are looking for a practical and understandable guide on the subject of electrical engineering. No matter what age you are, what profession you have, whether you are a pupil, student or pensioner. This book is for anyone who wants or needs to learn about electrical engineering and electronics.

The aim of this book is to introduce you to how electrical engineering accompanies us in everyday life and what basic principles are involved. It is a book that provides an understanding of electrical circuits and also an understanding of the most important components (e.g. resistor, transformer, capacitor, diode, etc.) in electrical engineering and electronics. In addition, you will learn the basics of direct current technology and alternating current technology, their physical backgrounds and much more! Develop a basic understanding of electrical engineering and electronics!

In this electrical engineering basic course you will learn everything you need to know as a beginner about the world of electrical engineering and electronics! So don't hesitate any longer, take a look at the book and get your copy as an ebook or paperback!

Table of Contents

1 Introduction

What to expect and what you will learn in this book

In this electrical engineering beginner's guide you will find an introduction to the basics of electrical engineering and electronics and learn in particular the basic terms and quantities such as current, voltage, power and the structure and application of important electronic components such as resistor, diode, transistor, capacitor and much more in detail. Step by step, I share with you as an engineer my knowledge from study and practice, so that you can achieve optimal learning success on the one hand with theoretical basics, but on the other hand especially with practical examples.

In this course, which is specifically aimed at beginners, you will also learn how electrical circuits are constructed and how they can be analyzed or solved. For this purpose we will use e.g. Kirchhoff's rules, which we will get to know in detail. In sample examples we will also do some calculations together and we will also learn the mathematical equations behind the basic principles of electrical engineering in each chapter. However, depending on how deep you want to go into the subject, you can also just take note of them. This book, besides equations, mainly provides an easy and understandable way to get started with electrical engineering and become more familiar with current & voltage with each chapter.

In brief, this course will teach you the following in detail:

- *Basic terms and basic quantities of electrical engineering*

- *Analyze and solve electrical circuits*

- *Ohm's law, Ampere's law and Farady's law*

- *Components such as resistor, diode (e.g. LED), transistor, capacitor, transformer, ... and get to know how they work and their areas of application*

- *The difference between direct current and alternating current, as well as single-phase and multi-phase systems (keyword: heavy current)*

- *How does electricity get into the house? Getting to know the power supply system*

- *Direct current and alternating current motors and their construction / mode of operation*

- *and much more!*

Be excited! Here we go!

2 Electrotechnical Basics & Circuit Analysis

2.1 Introduction to electrical engineering

Electrical engineering is largely based on two fundamental physical quantities, which are already covered in school - namely, charge and energy (work). Andre Ampere was the first to discover these properties of electricity, which are used in the form of current and voltage for the analysis of electrical and electronic circuits. It is important to distinguish between these two quantities. Without going into detail about the relatively complex quantum principles behind the physical nature of electric charge and energy (work), we will take this quantum nature for granted in this book and focus our attention more on practical applications. We will first deal with the two fundamental quantities of charge and energy, as well as the - often misunderstood - difference between current and voltage, before learning about Ohm's Law. The first few chapters in particular will be a little drier, as they are theoretical basics that are necessary for the rest of the chapters, so hang in there!

2.2 Basic sizes

The two basic quantities in electrical engineering are, as already mentioned: charge and energy.

Charge, measured in Coulombs (C) and described by the letter Q (or q), is a physical quantity that has the property of experiencing a force when placed in an electromagnetic field. What does this mean and what is an electromagnetic field? An electromagnetic field is composed of an electric field and a magnetic field, which are coupled together. It is a kind of state of space or an area where accelerated charges are present. Humans cannot perceive electromagnetic fields with their sensory organs in a differentiated way, with the exception of the visible range, which everyone perceives as light. Nowadays, it is hard to imagine life without electromagnetic fields. Every microwave works with microwaves of the same name, and every mobile phone also works with microwave radiation. But more about that later. There are two types of charges: The positive (+) and the negative (-). Equal charges repel each other, unequal charges attract each other. We come into contact with charges in our everyday lives more often than we would think. Who doesn't know the crackling and the disheveled hair when you put on or take off grandma's woolen sweater. Or the small electric shock when touching a door handle or a metal part, if the combination between shoe sole and floor covering (e.g. rubber sole and carpet) is unfavorable. The origin for these everyday experiences are charges. Every object has positive and negative charges that are normally in balance. However, friction during dressing or walking shifts this balance of charges, creating electrical voltage. When the hairs become charged as you put on your wool sweater, they either get stuck somewhere or appear to float as they repel each other. This happens because of the equal or opposite charge (two equal charges repel each other, two different charges attract each other).

Figure 1: Two unequal charges repel each other

However, charge is not suitable as a quantity for the analysis of circuits. For this purpose, we rather need the "current".

In electricity, the **current** (unit: ampere =. $\frac{\text{transported load}}{\text{time unit}}$; letter I or i), is defined as charges in motion and is therefore a more practical quantity. Charges in motion, simply put, are charges that are moved or transported per unit time, so we can express current mathematically as:

$$I(t) = \frac{\mathrm{dQ(t)}}{\mathrm{dt}} = \frac{\mathrm{C}}{\mathrm{s}} = \mathrm{Ampere} \qquad 1\text{-}1$$

In this book, we will use an *italic* symbol for scalar quantities and indicate vectors in **bold**. A scalar (e.g. mass, temperature,...) is simply a quantity characterized by the specification of a numerical value. A vector on the other hand (e.g. velocity) is a quantity described by a numerical value, a unit and a direction. Although current has a direction here, it is still not a vector quantity, but a scalar. Mathematically simplified, we can say that current addition (simple addition of particles, e.g. 3 + 4 charges = 7 charges) does not follow the laws of vector addition, so it cannot be a vector quantity.

Another known quantity, the **voltage** (unit: volt = $\frac{J}{s}$; letter U or v) can be understood as the energy change (or work) of a charge in motion. So if a charge of 1 coulomb experiences a change in energy of 1 joule, this means that there is a change in energy of 1 volt. We also call this **potential difference.** This potential difference (or voltage) can be described mathematically in electricity as follows:

$$U(t) = \frac{dW}{dQ} = \frac{J}{Q} = Volt \qquad 1\text{-}2$$

Now, the voltage does not depend on the movement of the charges (i.e. the current). We know this because charges cannot flow without energy. But at the same time, energy can exist without causing those charges to flow. How can we understand this?

For example, just imagine something that is so heavy that you cannot lift it. Even though you can't lift it, you are already applying energy by just trying. I.e. energy exists, but there is no movement. Similarly, in electricity, so-called insulators and open circuits have voltages, but no current can flow through them. It is also important to understand that voltage (U) does not depend on time (t) (see 1-2), whereas current (I) does depend on time (t) (see 1-1).

In summary, it can be said: charges in motion (current) requires energy

In addition, the following fact applies: If the value or quantity ($q = n \cdot e$) of a charge (coulomb) increases, more energy is needed for the charges to travel the same distance in one second. In above relation, 'n' is the number of particles and 'e' is the charge of an electron. So as the number increases, the value of the charge increases and more energy is required to travel the same distance in the same time.

2.3 Power equation and Ohm's law

A more general quantity, **power** (unit: watts), is defined as work per unit of time and is generally more practical to use because it includes time. Thus, power is the work done on a charge in a unit of time. In other words, we can also define it as the energy of a given number of charges (n-charges) in motion (current):

$$P = \frac{dW}{dt} = \frac{dQ}{dt} \cdot \frac{dW}{dQ} = U \cdot \mathrm{I} \qquad 1\text{-}3$$

We measure electrical power (P) in J/s, or $V \cdot A$ which is synonymous with the unit Watt, named after the Scottish explorer James Watt.

The unit of power, like many other units (SI units), is thus defined with the time unit second. In our everyday life, however, the time unit hour is often more practical. Therefore, the unit used for **energy in** everyday life is: kWh, which is 1000 watts times 1 hour. It can be seen as power (generation/consumption) for one hour. So 1 kWh is the energy that an appliance with a power of 1,000 watts takes in or gives off in one hour. To put it even more simply: If a light bulb of 20 W, runs continuously for 50 hours, it consumes an energy of 1 kWh ($20 \cdot 50 = 1000$). For this bulb, 20 W means the energy consumption of 20 J in 1 second, and 1 kWh just means the consumption of 20 W of power for 50 hours.

The interaction of voltage and power is an important relationship in the analysis of electrical circuits. There is another important relationship that we will briefly look at below, so that we can solve any circuit problem with the combination of these two relationships.

Imagine current (I) flowing in a conductor under the influence of voltage (U). In this scenario, the charge particles of this current (I) collide with each other and sometimes with the walls of the conductor. This collision of charges develops a ***resistance R in*** their flow (measured in ohms or Ω), and as this resistance increases, the charges slow down (current decay). Since the voltage U is directly proportional to the current I, the mathematical definition is thus:

$$V \propto I \Rightarrow U = R \cdot I \qquad 1\text{-}4$$

"R" is the constant of proportionality of current and voltage, "∝" means directly proportional.

Now let's relate this equation, which is also known as Ohm's Law, to the power equation. For example, simply consider a 200W light bulb compared to a 100W light bulb. Since the 200-W bulb logically (200W > 100W) has more power, more current will flow through it, and from (1-4) we get a lower resistance (1-5). The arrows stand for this in the following for an amplification or attenuation of the individual quantities.

$$P \uparrow = U \cdot I \uparrow \text{ and } R \downarrow = \frac{U}{I\uparrow} \qquad 1\text{-}5$$

This can be a bit confusing at first, as it is hard to imagine how increasing the current can decrease the resistance. Normally, you might think that if the current is increased, more particles **should** collide with each other and therefore there should be more resistance. However, this is not the case! Here, with the help of a little thinking about how current is directly related to power and inversely related to resistance, you can get a good understanding of this.

Before we get into the first circuit, let's learn about another quantity, **conductance.** The conductance is the reciprocal of the resistance. ($G = \frac{1}{R}$). Conductance is used to get an idea of the electrical conductivity of a material. Conductivity and resistivity are umbrella terms because different materials have different capabilities. The **resistivity value** (ρ ; *pronounced rho*) and the **conductivity** ($\frac{1}{\rho}$) are often used in practical cases when the resistance of a particular material is needed. The following equation relates the resistance to the resistivity value:

$$R = \rho \cdot \frac{L}{A} \qquad 1\text{-}6$$

This equation (1-6) basically just states that the resistance of a component depends on the resistivity value (ρ), which is defined for each material in terms of a fixed value, and the length of the conductor as well as the area of the conductor cross-section (e.g. cable cross-section).

Using equations 1-3 and 1-4, we can derive even more relationships using power, relating it to resistance. For example:

$$P = U \cdot I = U \cdot \frac{U}{R} = \frac{U^2}{R} = \frac{I^2 \cdot R^2}{R} = I^2 \cdot R \quad (1\text{-}7)$$

What is a circuit? Simply put, a circuit is an arrangement of different components with an electrically conductive connection between them. In order for an electrical circuit or circuit to work, you need a power source / current source, such as a battery and a load, such as a light bulb, as well as connections between these two components, which are called conductors. In electrical engineering, these components are represented as symbols in a circuit or a circuit.

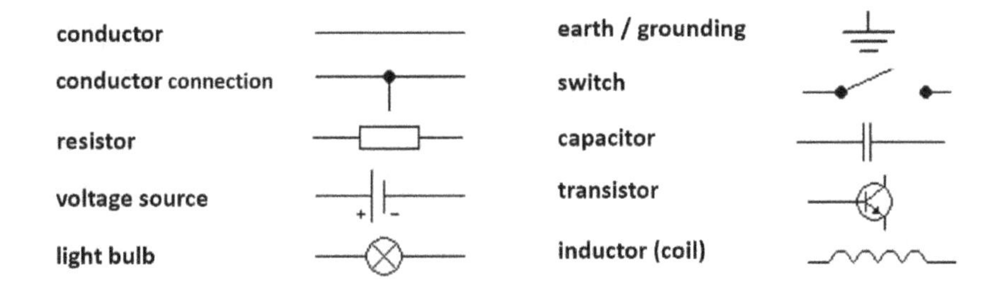

Figure 2: The most important circuit symbols

In order for a lamp, for example, to light up as shown in Figure 3, the **circuit must be closed, i.e.** there must be a connection between the two poles (+ and -) of a power source (e.g. battery) and the light bulb. If this is the case, current flows from one pole of the power source (e.g. battery) through the bulb and back to the other pole of the power source. When this connection is severed, e.g. by a switch, current no longer flows and the bulb no longer lights. In this case, it is called an **open circuit**. A **short circuit** occurs if the current can flow unhindered and without first passing through an electrical component from one pole of the current source to the other pole (e.g. through an uninsulated spot of a cable on a metal surface). This is because the current always takes the path of least resistance.

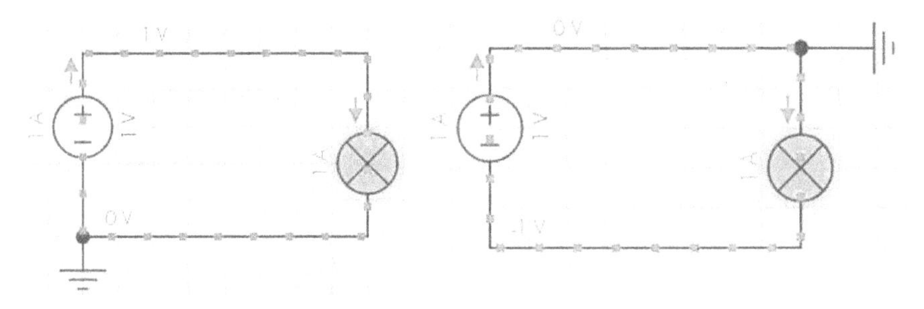

Figure 3: Circuits with battery (+ / - symbol) and lamp (yellow)

Sample Example 1
What is the ratio of resistance to amperage for a 100 W lamp operating at 220 V compared to a lamp of the same wattage operating at 110 V?

There, P_1 and P_2 are equal:

$$P = \frac{U_1^2}{R_1} \Rightarrow R_1 = \frac{U_1^2}{P} R_2 = \frac{U_2^2}{P}$$

$$\frac{R_1}{R_2} = \frac{220\,V^2}{110\,V^2} = \frac{1}{4} \qquad \boxed{\Rightarrow R_2 = 4R_1}$$

So if the voltage is halved, the resistance increases by 4 times. Using Ohm's law, you get:

$$\Rightarrow \frac{U_2}{I_2} = 4\frac{U_1}{I_1}$$

$$\Rightarrow \frac{110\,V/_4}{220\,V} = \frac{I_1}{I_2} = \frac{1}{8} \qquad \boxed{\Rightarrow I_2 = 8I_1}$$

So the current increases 8 times.

In the following, we will see how equation 1-3 can be used in the analysis of circuits and the so-called Kirchhoff's laws and then discuss application-related problems of the series-parallel circuit, which are 1-based on equation 1-4 . These two equations, as we will see below, are the basis of circuit analysis. That is why it was so important to discuss these equations in this first section. If you have not yet fully understood these basic equations, it is best to review the first section briefly so that you can develop a good understanding for solving the problems that follow.

2.4 The passive sign convention

Before we get into some practical examples of electronic circuits and analyze them, we will first look at signs and the so-called sign convention in this chapter. In circuit analysis, electrical engineers use a sign convention (passive or active) to match signs (i.e. + or -) in calculations.

The **passive sign convention** is the most commonly used convention, so we will use it throughout this book for circuit analysis. The passive sign convention simply states that **the power of passive elements (i.e., components that draw power - e.g., lamps or motors) is positive ("+" as the sign) and of active elements (i.e., components that dissipate power - e.g., a battery or a discharging capacitor) is negative ("-" as the sign).** This means for equation 1-3 ($P = U \cdot I$)that a device draws power when the signs of the voltage and current match ("+" times "+" equals "+"; "-" times "-" also equals "+"). If the signs of voltage and current are different ("+" times "-" results in "-"; applies to both directions), a device is outputting power.

The positive and negative signs of voltage and current mean that they match or do not match the **reference direction**. That is, whether the current and voltage are moving in the direction of the reference or away from it. The reference is often referred to as **ground** ⏚- see Appendix-A for other symbols) and is often seen as the electrical negative pole at the same time. The reference is like an origin (with a value of zero) and is used as a starting point in calculations. To illustrate this, let's look at a small example using Figure 3 from the previous chapter.

In Figure 3 (we are looking at the left side), for the voltage source represented by the battery symbol, the current is *negative* because it flows away from the reference (ground⏚ ; see green arrows), i.e. in the negative direction. Why does the current flow in this direction? Because the technical current always flows from "+" to "-". The voltage, on the other hand, is *positive because the* negative sign of the voltage source corresponds to the reference, and thus the power P (= $+U \cdot -I$) is negative.

In the same way, for the lamp (a passive element), the voltage is *positive (coincides* with the reference) and the current is also *positive* (moves towards the reference), and therefore the power P (=$+U \cdot +I$) is positive. The lamp draws power according to convention. The passive sign convention is quite intuitive and that is why it is often used. In very simple terms, here the conventional current is negative when it moves from the negative to the positive side of the battery, which means that work is being done on it.

2.5 Analysis of direct current circuits

In this section we will cover some commonly used methods (Kirchhoff's laws, mesh current analysis, nodal voltage analysis) for solving circuits. Solving circuits means calculating the unknown and desired parameters like voltages and currents from already known / given values. We will first start with the circuit terms and then learn the individual laws and methods in an intuitive manner. In this section, we first deal with **direct current** (DC). There is also **alternating current (**AC). The difference between direct current and alternating current is basically that direct current always flows in the same direction. The direction of alternating current, on the other hand, changes, as we will see in more detail in one of the next chapters. By the way, there is also the so-called **mixed current,** which results from a direct current and an alternating current component, i.e. a superposition. However, we will not deal with this here.

2.5.1 Terms in circuits

Electrical components can be thought of as a type of small device (such as a battery or capacitor) that is treated as a separate entity from the rest of the circuit. Each element actually has some resistance of its own due to the material and wires it contains. However, we use **idealized elements (lumped elements)** when analyzing circuits. An idealized element is simply a component consisting of two terminals that are so short that the resistance through the lead wire is zero.

In an electrical circuit, a **node is** a point that separates two electrical elements. There are **meshes or loops** where the current starts from a point, follows a circular path and returns. In Figure 4 (next chapter), points "A", "B", "C", and "G" (ground) represent nodes. In the same way, the currents "A" and "B" (blue) represent two loops.

An element can be **linear or non-linear**. An element is linear if it follows Ohm's law, i.e. the current-(I)-voltage-(U)-relationship (e.g. simple resistor). Semiconductor devices like transistors, which we will get to know in one of the next chapters, do not follow Ohm's law and therefore fall into the category of non-linear elements. In this chapter, however, we will only analyze circuits with linear elements.

An electrical circuit consists of **passive and active elements**, which are defined as devices that consume (e.g. lamp) or emit (e.g. battery) electricity. We have already learned this in the passive character convention. Some active elements also depend on other current sources and are therefore called **dependent sources. Dependent** sources are usually used for amplifier analysis and therefore work with a gain term multiplied by the current/voltage of an independent source. In the simple analysis, we use only **independent sources** and treat the passive elements using the equations we have already learned.

2.5.2 Kirchhoff's laws (KCL & KVL)

For the analysis of circuits one makes use of Kirchhoff's laws, which we will get to know in this chapter. As we already know, in an electrical conductor the current is simply the flow of charges. Now**, Kirchhoff's law of current (KCL)** is simply defined as follows: **The sum of all currents flowing in a node is always zero.** How can we apply this law to our circuit from figure 4? With the help of equation 1-8, which we can easily set up for node "A" as follows: A current I_{CA} flows from the node "C" towards the node "A", therefore we write $+I_{CA}$ and two currents I_{AG} & I_{AB} move away from the node "A", therefore we add $-I_{AG}$ as well as $-I_{AB}$. Now as you may have noticed, in this book we will add positive sign to the currents moving towards the nodes and we will add negative sign to the currents moving away from a node.

$$I_{CA} - I_{AG} - \mathrm{I}_{\mathrm{AB}} = 0 \Rightarrow I_{CA} = I_{AG} + I_{AB} \qquad \text{1-8}$$

Furthermore, if we rearrange equations 1-8, we can see that the current I_{CA} results from the addition of the other two currents. So to solve this circuit we only need the values of these currents.

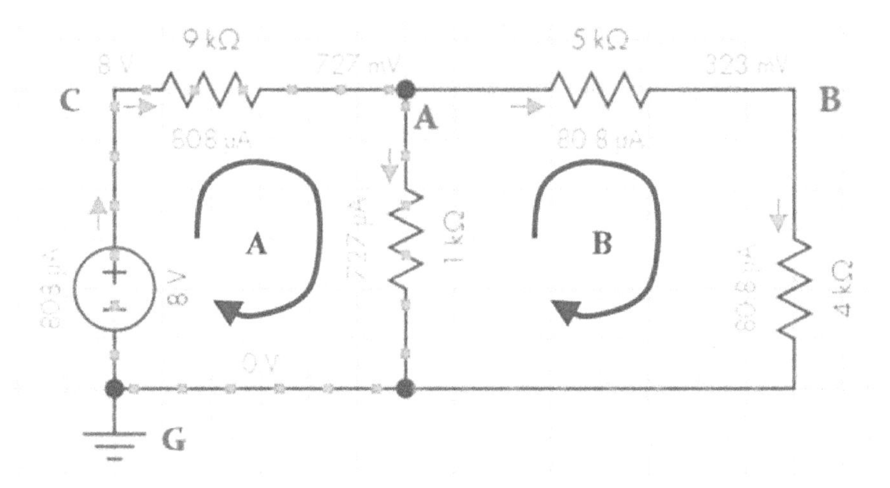

Figure 4: A circuit with one power supply, four resistors, two loops (blue) and 3+1 nodes (red)

Kirchhoff's Voltage Law (KVL) is relatively similar to the Current Law, it is simply defined as: **The sum of the voltages in a circuit is always zero**. This means that in any circuit, the voltage of the passive elements always balances with the voltage of the active elements, so that their sum is always zero. According to the passive sign convention, the voltage of the current source is defined as negative. So, expressed in an equation, this voltage law is for the circuit in Fig. 4:

$$U_{CA} + U_{AB} + U_{AG} + U_{BG} - U_{CG} = 0 \qquad \text{1-9}$$
$$\Rightarrow U_{CA} + U_{AB} + U_{AG} + U_{BG} = U_{CG}$$

2.5.3 The mesh flow analysis

Mesh current analysis is a method that uses Kirchhoff's Voltage Law (KVL), which we learned earlier, to solve for variables in a circuit. In our example, if the unknown loop currents
("A" and "B" in blue - Figure 4) are multiplied by resistances (keyword: Ohm's law), the sum of the resulting voltages of all elements equals zero. Let's have a look at this by means of a calculation of our example:

Sample Example
Use the mesh current method to determine all of the unknown variables in the circuit in Figure 4 given only the source voltage (8 V) and resistors (1, 9, 5, and 4 kΩ).

Loop A:

There are three elements here with two currents ($I_A$and I_B) flowing through a 1 kΩ resistor. For loop A, Adue to the passive sign convention in this loop, I is positive while I is Bnegative, and we can set up the KVL equation as follows:

$$\mathbf{-8\,V + I_A \cdot 9\,k\Omega + I_A \cdot 1\,k\Omega - I_B \cdot 1\,k\Omega = 0}$$

$$\Rightarrow 10\,k\Omega \cdot I_A - 1\,k\Omega \cdot I_B = 8\,V \qquad (1)$$

Loop B:

$$I_B \cdot 1\,k\Omega - I_A \cdot 1\,k\Omega + I_B \cdot 5\,k\Omega + I_B \cdot 4\,k\Omega = 0$$
$$\Rightarrow 10\,k\Omega \cdot I_B - 1\,k\Omega \cdot I_A = 0 \qquad (2)$$

Release (1) with (2):

$$100\,k\Omega \cdot I_A = 8\,V \Rightarrow I_A = 8\,V\,/\,100\,mA =$$
$$0,808\,mA = 808\,\mu A$$
$$\Rightarrow U_{CA} = 808\,\mu A \cdot 9\,k\Omega = 7,272\,V$$

and with (2)

$$I_B = \frac{808 \cdot 10^{-6} \cdot 1000}{10 \cdot 1000} = 80,8\,\mu A$$
$$\Rightarrow U_{AB} = 80,8\,\mu A \cdot 5\,k\Omega = 0,404\,V$$
$$\Rightarrow U_{BG} = 80,8\,\mu A \cdot 4\,k\Omega = 0,3232\,V$$

The voltage in the middle resistor is:

$$I_{AG} = I_1 - I_2 = (808 - 80,8)\,\mu A = 727,2\,\mu A$$
$$\Rightarrow U_{AG} = 727,2\,\mu \cdot 1\,k\Omega = 0,7272\,V$$

Note that the sum of all the currents of a node is also zero. This helps us to check our results.

2.5.4 The nodal stress analysis (nodal analysis)

We can see in the above example that an element voltage can be equal to the difference of its connecting nodes. For example, U_{CA} is equal to $U_C - U_A = 8\,V - 0,7272\,V = 7,272\,V$ (see also Fig. 4). This statement forms the basis of the **node stress analysis**. In our case, we can write generalized:

$$U_{CA} = U_{CG} - U_{AG} = U_G - U_A \qquad 1\text{-}10$$

In node voltage analysis, unlike the mesh current analysis from earlier, we solve circuits using Kirchhoff's Current Law (KCL). When we determine the currents flowing in each node, we obtain node equations, which then give us the desired results once they are solved. For the circuit in Figure 4, there are three (4 - 1 = 3) independent equations to solve. There are 4 nodes, but we exclude the ground node "G" since it is not an independent equation. These three equations are sufficient to solve for the three unknown variables: U_A, U_B and U_C. These equations for the circuit in Figure 4 are:

Node A: $I_{CA} - I_{AG} - I_{AB} = 0$
Node B: $I_{AB} - I_{BG} = 0$
Node C: $I_{GC} - I_{CA} = 0$

As you can see, we have used the passive sign convention here for the signs of the flows. Currents moving towards the references (Node A, Node B and Node C) are positive and currents moving away from the references are negative. Finally, we can use Ohm's law $I = \frac{U}{R}$ and Equation 1-10 to solve for the voltage across the three nodes. However, we will see how to do it more simply in the next chapter!

2.5.5 Equivalent circuits

Using the two simple rules we learned earlier, we can analyze any electrical circuit. But sometimes it can be tedious to solve an equation for each circuit, as is the case with our example in Figure 4. Here it is much easier to use equivalent circuits. Equivalent circuits reduce complex circuits into a simple form to make calculations easier. For example, the circuit in Figure 4, can be reduced to a single voltage source and resistor. To do this, we use what are called series and parallel circuits.

A **series** circuit is a circuit in which two elements have a common node or, more simply, are arranged in series, while in a **parallel circuit each element** has two independent connection nodes or, more simply, the elements are arranged in parallel. Figure 4 shows resistors connected in series and parallel. Sometimes it can be difficult to judge whether the elements are in series or parallel. However, the above definitions based on nodes can help us deal with such situations.

2.5.6 Properties of series and parallel circuits

If we imagine resistors connected in series or also in series, each resistor would impede the flow of current as the current flows through it. So the total resistance or **equivalent resistance of** the individual resistors in a series circuit is obtained by adding the individual resistors together. In a parallel circuit of resistors, simply put, the current splits in each path. The sum of the conductances (the conductance is the reciprocal of a resistor, i.e. 1/R) of the individual paths is therefore equal to the total conductance.

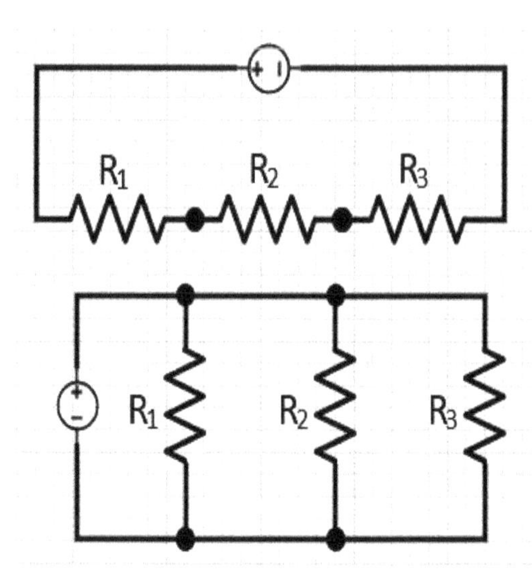

Figure 5: Resistors connected in series (top) and in parallel (bottom)

SERIES:

$$R_{gesamt} = R_1 + R_2 + R_3 \quad (1\text{-}11)$$

PARALLEL:

$$G_{gesamt} = G_1 + G_2 + G_3 \quad (1\text{-}12)$$

$$\Rightarrow \frac{1}{R_{gesamt}} = \frac{1}{R_1} + \frac{1}{R_2} + \frac{1}{R_3}$$

In our example from Figure 4, the resistors with 5 kΩ and 4 kΩ (right area) are connected in series and this series connection is connected in parallel with the 1 kΩ resistor (middle section). In addition, a 9 kΩ resistor (upper left) is connected in series again. With these formulas we can reduce the circuit in figure 4 as follows (Note: || stands for parallel):

$$((5k\Omega + 4k\Omega)||(1k\Omega)) + 9k\Omega = 9k\Omega||1k\Omega + 9k\Omega$$

$$\Rightarrow \frac{1}{9k\Omega} + \frac{1}{1k\Omega} + 9k\Omega = 9k\Omega$$

By multiplying equations 1-11 & 1-12 by the voltage we can see that **in the series circuit the voltage adds up, while in the parallel circuit the current adds up.** This is easy to understand if we consider that each resistor reduces the energy of the current and the current in a parallel circuit takes different paths, but the sum of these must again correspond to the total current, ultimately the current only splits, but does not become more. Since in the parallel circuit the conductance of the path with the lowest resistance is the highest, more current flows through this path. If all three resistances are equal, an equal amount of current flows through all three. Generally, when analyzing a circuit, as in the parallel circuit, it is more practical to simply apply equations 1-11 & 1-12 rather than performing the node or network analysis. To do this, redraw the circuits and use Ohm's law for the unknowns. Let's look at the following example for clarification:

Sample Example 3

Calculate the equivalent resistance when the value of each resistor in the cube of the figure below is "R".

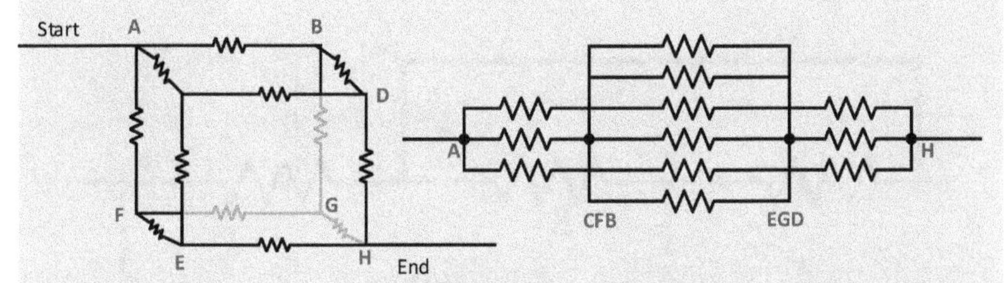

We can convert this cube into an equivalent circuit as shown on the right side of the figure. Now we can see that the resistors are connected in series and in parallel. Therefore, we can now set up the corresponding equations for the equivalent resistance.

From A-(CFB):

$$\frac{1}{R_{eq_1}} = \frac{1}{R} + \frac{1}{R} + \frac{1}{R}$$

$$\Rightarrow R_{eq_1} = \frac{R}{3}$$

Note that for 'n' resistors connected in parallel, the equivalent is equal to R/n since the total current is distributed equally to each path

From CFB-EGD:

$$R_{eq_2} = \frac{R}{6}$$

From EDG-H:

$$R_{eq_3} = \frac{R}{3}$$

So we see that they are all connected in series:

R_{eq_1} R_{eq_2} R_{eq_3}

CFB EGD H

Results:

$$R_{eq} = R_{eq_1} + R_{eq_2} + R_{eq_3} = \frac{2R}{3} + \frac{R}{6} = \boxed{\frac{5R}{6}}$$

Now that the circuit in the above example has been converted to a single equivalent resistor, we can apply Ohm's law ($U = R \cdot I$ or rewritten $R = \frac{U}{I}$ or $I = \frac{U}{R}$) to determine voltage or current. For example, if the connected power supply is 12 V, then the current is $I = \frac{12V}{\frac{5R}{6}} = \frac{12V}{5R} \cdot 6$. "R" here is a fictitious placeholder for a resistor value.

This method is quite simple, but sometimes it can be difficult to redraw a circuit. As an exercise, consider how the resistors in the circuit in Figure 6 are connected.

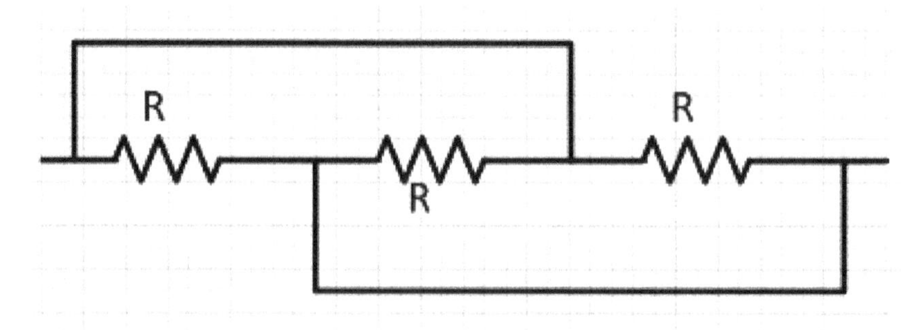

Figure 6: How are these resistors connected?

We cannot solve all problems with equivalent or substitute resistors. Therefore, it is important to know the KCL and KVL analysis techniques already presented. In addition, there are other theorems or even relationships that engineers have found to facilitate problem solving. In this section, we would like to learn about the **voltage and current divider rule** before moving on to the next chapter, electronics. By the way, in addition to the voltage and current divider rule, there are also the Thevinen, Norton, and superposition theorems, which are also commonly used in analysis. However, these are outside the scope of this beginner's book and will therefore not be covered.

To learn the concept of voltage and current divider rule, consider the following problem: The voltage divides in a series circuit because the resistors impede the energy of the flowing charges, but by how much does it divide when the resistors have different values? For a simple series with resistors R_1 and R_2 and source voltage U_S, we can solve this problem with KVL as follows:

$$R_{total} = R_1 + R_2$$

$$I = \frac{U_s}{R_1 + R_2} = \frac{U_{R_2}}{R_2}$$

Voltage divider

$$\Rightarrow U_{R_2} = U_s\left(\frac{R_2}{R_1 + R_2}\right)$$

$$U_{R_1} = U_s\left(\frac{R_1}{R_1 + R_2}\right) \qquad 1\text{-}13$$

Similarly, current divides in parallel, and for the current divider we can use KCL at a node of two resistors $R_1 || R_2$, with current I_S. The result is:

Current divider

$$\Rightarrow I_{R_1} = I_s(\frac{R_2}{R_1 + R_2})$$
$$I_{R_2} = I_s(\frac{R_1}{R_1 + R_2}) \qquad 1\text{-}14$$

3 Basics of electronics

In the previous chapter, we looked at the basics, the names and relationships of electrical systems. We discussed basic circuit analysis techniques, such as Kirchhoff's laws or mesh and node rules. This involved active circuit elements that control variables (such as current) through passive elements. In this chapter, we will look at the basics of electronics, a major area of electrical engineering.

Humans have been trying to develop tools to help them survive since the beginning. In this day and age, we can easily buy everything we need to survive. Mere survival has taken a back seat and the desire to live as comfortable and happy a life as possible has come to the forefront. Meanwhile, our society has reached the point where we even have machines to do repetitive tasks in an automated way. Think of your washing machine or a dishwasher, for example. From a simple electrical switch to a complex communication device (cell phone), our electronics have evolved. Electronics is also about controlling things, which in descriptive terms we can do by manipulating the flow of electrons through various devices in analog and digital ways.

In the last chapter we controlled current by passive elements. For example, if we need 5 V from a 12 V battery, we simply connect two resistors of 1 kΩ and 715 Ω in series. By now we can calculate this quite easily with equation 1-13.

In analog electronics, we use passive elements for control. However, it can be difficult - depending on the requirement - to design circuits in an analog way only. Here it is an advantage that an equivalent electronic device of an analog circuit can also be designed in a digital way. The basis of digital electronics is simple switching operations. The computer is one of the best examples of these switching operations and of digital electronics. The applications that a modern computer gives us are achieved using switching operations performed by millions of transistors. In this chapter, we will deal mainly with such switching devices. However, we will also briefly discuss capacitors and inductors for analog circuit design in this chapter.

The invention of the radio in the late 19th century is generally considered the beginning of the electronic age. The technology of a radio used electromagnetic waves for the first time in a very special way, namely for communication. Later, after the invention of transistors in 1947, the era of digitally controlled applications (e.g. computers) emerged. In today's world, electronics continue to expand - every year more compact circuits with higher computing power and efficiency are developed, leading to a more controlled and automated world. There are now countless applications of electronics and we find them everywhere, in our homes, on the streets and in offices. Many electronic components combine to create a wide variety of devices, such as mobile phones, tablets, televisions but also simple street lights. It is hard to imagine today's technological world without electronics.

In this chapter we will first cover the basics of electronics and then get an introduction to specific components such as diodes, transistors and more.

3.1 Fundamentals of semiconductors

The following basics may go into a bit more detail and may be a bit more difficult to understand when reading for the first time and depending on previous knowledge. Nevertheless, it is useful to have heard the important terms (in bold) in connection with semiconductors at least once. There is no shame if you don't understand all or very little on first reading. Just read these sections two or three times and stay tuned, more practical examples will follow in the further chapters.

In quantum mechanics, electrons in an atom, distributed in **shells,** are described. The shells are classified by a number called the **principal quantum number.** Each shell has its **subshells**, in which the electrons have the same **principal quantum number** (again describing the subshells). Each element in the periodic table has a different atomic number and therefore a different number of electrons. Each electron in an atom has its own energy state. When an electron is excited, it moves to a higher energy shell of the atom (**Bohr's atomic model** and **energy band theory**). The electrons of the outermost occupied shell (valence shell) have the highest energy compared to the other shells. When the temperature rises or a certain potential is applied to an element, the electrons in the valence band become excited and move to a higher level/conduction band. This is where the current begins to flow. The valence electrons of different elements require different amounts of energy to be excited out of the valence band.

Those atoms with fewer bonds in their valence shell are the more conductive. To some extent, this also depends on properties such as ionization energy, electronegativity and atomic radius of the elements. Elements in group 11 of the periodic table have only 1 electron in their valence shell, so even though they have a high ionization energy and a small radius, they are more conductive. In group 11, silver is more conductive than copper because it has a large radius. Gold, on the other hand, is less conductive because of its high ionization energy (IE). Silicon (Si), for example, is used for semiconductors.

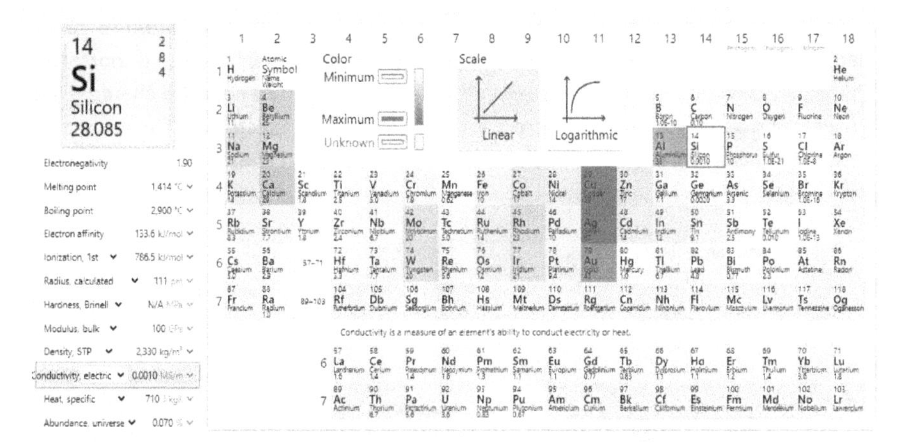

Figure 7: Properties of silicon (Si) (left)
Periodic table scaled with conductivity of the elements (right); (Source: https://ptable.com)

Semiconductor elements belong to the IVA group of the periodic table, so they have four electrons in their outermost shell. The valence electrons normally form four covalent bonds with neighboring silicon atoms, resulting in a crystal-like lattice as the ionization energy increases with period and decreases from top to bottom. Semiconductor elements lie between this trend. So unlike conductors and insulators, leaving the valence band is neither too easy nor too hard for semiconductors. That is, they can conduct electricity, but unlike conductors, semiconductor electrons do not create too much turbulence when a potential is applied. So instead of a gain in resistance, semiconductors increase in conductivity.

3.1.1 Semiconductor doping

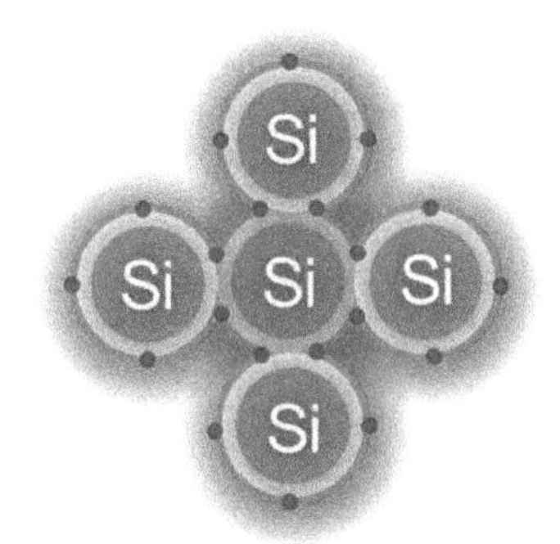

Whenever a valence shell electron (electron of the outermost occupied shell) of the silicon surface lattice leaves the valence band, a hole (positive charge) forms in its place in the atom. Now, when we apply an electric potential, the electrons flow from a low to a high potential, while the holes do just the opposite. In semiconductors, the total current is simply the sum of the electron and hole currents.

Intrinsic (pure / undoped) semiconductor elements have only a few electrons and therefore cannot release too many of them to the conduction band. The crystal lattice of silicon (see figure above right) is a perfectly bound structure, i.e. to release some electrons from the valence band into the conduction band, more energy is required. To get more free electrons or holes, **doping** of semiconductor elements is often used. We can get an extra free electron by bonding each element of Si with a group VA atom (5 valence electrons), such as phosphorus (called **donor atom also donor atom / donor atom**). This is called **n-type doping,** because in this case we get more negative charge. Figure 8 (right) shows n-type doping. Here we see an extra free electron left over from the phosphorus, leaving behind a positive ion. This electron acts as a free charge carrier and moves to where the potential takes it.

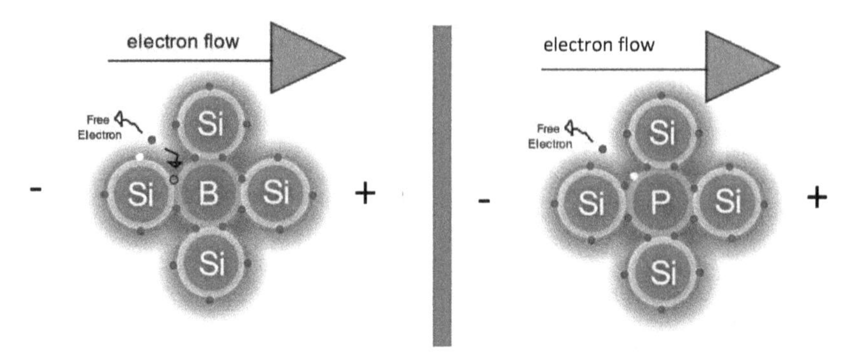

Figure 8: Atomic structure of Si under potential (voltage).
left: p-type doping - the free electron of the silicon fills the hole of the boron and lowers the conductivity right: n-type doping - the additional electron of the phosphor increases the conductivity.

If we now connect this silicon with an atom of group IIIA of the periodic table (3 valence electrons), such as boron (**acceptor atom / receiver atom**), more positive charge is generated. This is called **p-type doping.** Here, holes act as free charge carriers, since there is only a limited number of electrons.

Doped semiconductor elements are generally referred to as **extrinsic** (impure) semiconductors. An intrinsic semiconductor has the same number of electrons and holes ($n = p$) while n-type has a predominance of electrons and p-type has a predominance of holes. When we combine these two (n-type and p-type), the number of electrons and holes becomes equal again. This combination of donor and acceptor type semiconductors is called **PN junction.**

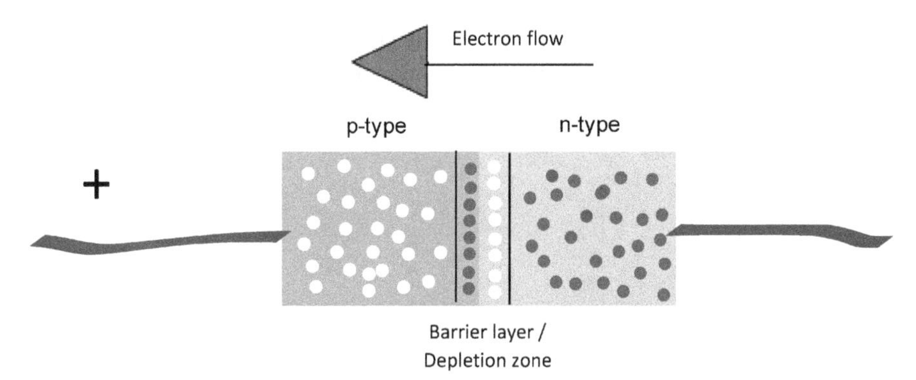

Figure 9: Diode (n-type and p-type doped semiconductors combined) connected to power supply. Free holes (white) & free electrons (blue)

3.1.2 The PN junction diode

The combination of these two types of semiconductors (p-type, n-type) has an important property. To make it easier to imagine, you can look at the current flow Figure 9. First, imagine free holes of donor atoms (p-type) moving towards the n-type electrons. As these gather in the n-type region, they repel more holes, which creates a resistance in the flow of holes toward free n-type electrons. In the same way, free n-type electrons also create this resistance as they flow from the n-type to the p-type. Once equilibrium is reached, a gradient called the **depletion zone / junction or space charge zone (RLZ) is** created, which stops the flow of electrons and holes between the regions.

In your imagination, now connect p-type / anode to the positive terminal of the battery and n-type / cathode to the negative terminal (as shown in Figure 9). As the potential increases, more n-type electrons enter the conduction band. When this mass of electrons flows towards the positive terminal of the battery, it breaks through the junction (depletion region) and the current flow in the circuit starts. This condition is referred to as **forward bias / forward direction ("forward bias").** Breaking through this depletion zone / junction for forward bias requires 0.7 V for silicon semiconductors and 0.3 V for germanium semiconductors.

Connecting the negative pole to the p-type and the positive pole to the n-type, on the other hand, never breaks through the barrier (junction), but makes it wider. In this case, no current can flow. This is called **reverse bias / reverse direction** (**"reverse bias").** When the bias is reversed, current never flows. Nevertheless, a large increase in voltage can break the entire junction at a point, which is called **breakdown voltage ("breakdown voltage"). The** forward and reverse curves are shown in Figure 10. Here on the x-axis (horizontal) is the voltage and on the y-axis (vertical) is the current. The ratio of current to voltage in forward gear (forward bias) and reverse gear (reverse bias) can therefore be read here.

This PN junction is commonly known as a **diode** and is often used in engineering for control because it **allows current flow in only one direction** (forward direction) and blocks it in the other.

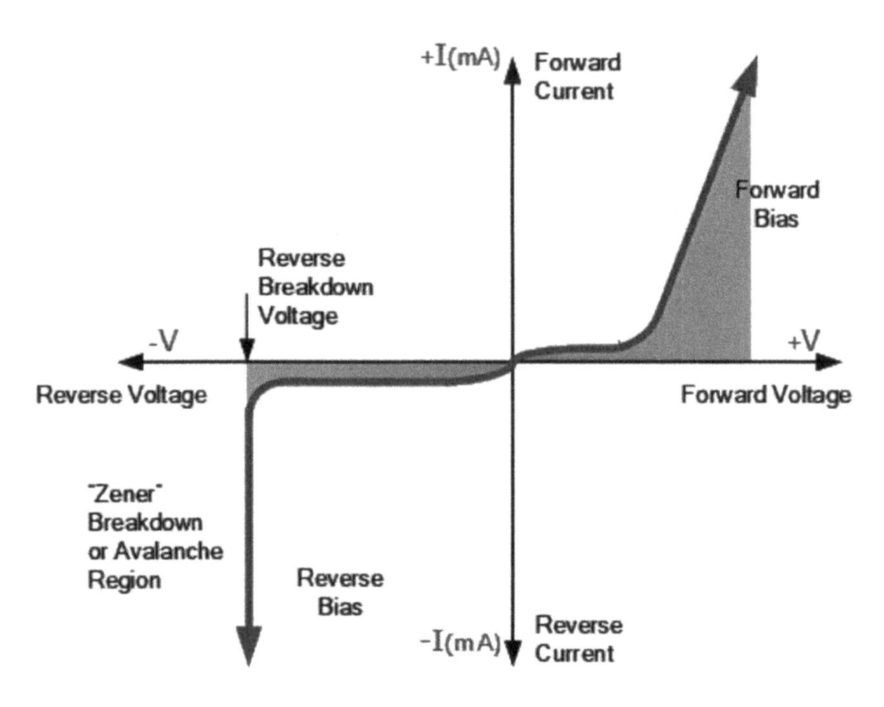

Figure 10: Forward and reverse characteristic curve

3.1.3 The light emitting diode (LED)

The simplest application of a diode is the LED. The LED (light-emitting diode) is a semiconductor device that produces light when it is energized. The light is produced by current flowing from a DC source to the diode and through it. Since an LED is a semiconductor device, it also has a forward direction. This means that current can only flow through it in that direction. If an LED is connected incorrectly, no light will be produced. The color of the light and whether it is visible or not (e.g., infrared; generally determined by wavelength) is controlled by the doping and material used. Two major advantages of LEDs are: a) long life, b) low power consumption. Compared to old-fashioned incandescent lamps, an LED can achieve a service life of several 10,000 hours and is many times more efficient. Why is this so? Conventional incandescent lamps produce an enormous amount of heat in addition to visible light, i.e. the energy expended is not only converted into light, but primarily into heat. With LEDs, only a little heat is produced as a "waste or by-product" and almost all the energy can be used to produce light. There are now several different types of LED. The simplest design is shown in Figure 11. The heart and also the actual semiconductor element of the LED shown is the LED chip, which is placed on a reflector on the anode and emits the light. The circuit symbol of an LED consists of the diode circuit symbol with two additional oblique arrows, which are intended to represent emitting light.

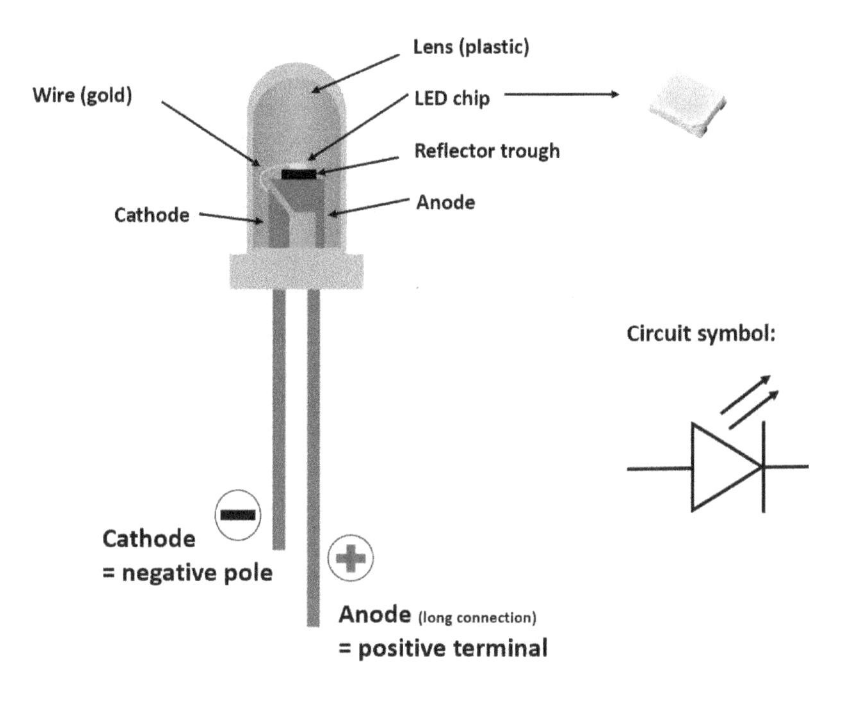

Figure 11: The structure of a simple LED and the circuit symbol of an LED

3.1.4 Solving circuits with diodes

To solve a circuit with diodes, we can now simply think of the diode as a circuit component and then use KCL and KVL (Kirchhoff's rules, see previous chapters) to solve for unknown variables. Just think of a diode as a device that absorbs a certain amount of voltage (volts) (in the case of silicon, this is 0.7 V), and apply KVL to the current. That is, the output voltage of a Si diode connected to a 5 volt supply will be 4.3 V. In complex cases, however, the solutions are not as simple as described in this beginner's book, because diodes have some other properties that are not explained in detail here but are mentioned briefly below. For example, there is a **saturation current, which** gives information about the reverse-biased charge flow, and a **junction capacitance,** which is defined as the capacitance that arises between the plates when they are reverse-biased. Parameters like these are important when simulating complex practical circuits with diodes.

3.2 Rectification and use of a diode

Besides LEDs, there are many other applications of a diode. Before we go into the details of important semiconductor devices such as bipolar transistors (BJT) and MOSFETs (metal oxide semiconductor field effect transistors), it is important to understand the role of the diode in the world of electronics: Diodes are most commonly used for rectification (rectification), which is the conversion of alternating current to direct current. However, the uses of a diode are not limited to rectification. We can also use them for temperature measurement, for sound reproduction in radios, as already known for light generation (LED) and many more. Diodes are one of the most basic elements in electronics and can be found in almost every application around us (e.g. diodes are also used in mobile phones to rectify modulated radio waves).

3.2.1 Half-wave rectifier

Now, as we already know, a diode only allows current to flow in one direction (the forward direction) and blocks the flow in the reverse direction. This happens because more charge carriers are added to this depletion zone in the backward direction (instead of breaking through the junction as in the forward direction) and this zone becomes wider as a result. So when an oscillating signal (vibration signal) passes through a diode, the diode returns only half of that input signal (the negative signal portion is clipped). This is called a half-wave rectifier. Let's look at a practical example of this:

When a digital square wave (see Figure 12, top) passes through a diode, the diode allows only the positive half to flow. Figure 12 (below) also shows the circuit and its associated waveform in the region before and after the diode. Here, we used a square wave generator with a frequency of 1 kHz and an amplitude between 1 V and -1 V. We can see this waveform in the purple wave (positive and negative components) of the figure. When this wave passes through the diode, the diode cancels the negative values and outputs the green wave (only positive parts from 0 V to 1 V).

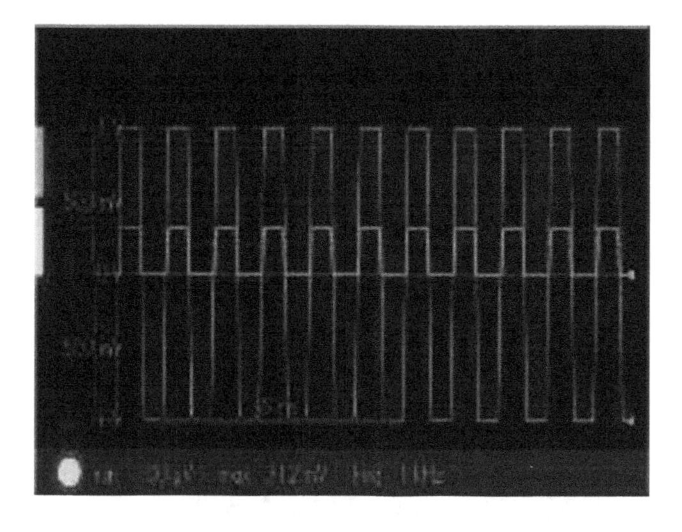

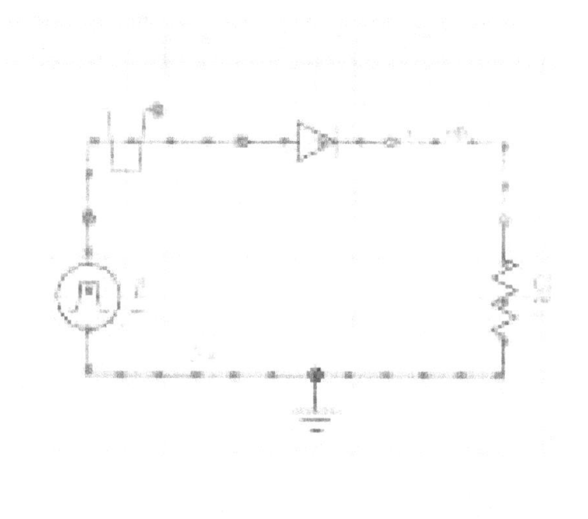

Figure 12: Half-wave rectifier. The purple square wave is the input and the green square wave is the output

3.2.2 Full wave rectifier

It is possible to convert the capped negative half of the single-wave rectifier just presented into a positive half. This is called full-wave rectification. For this purpose, a circuit with four diodes connected together in a bridge, hence sometimes called a **(diode) bridge rectifier,** is used. Figure 13 shows the schematic (below) of this circuit and the output signal (above). The purple wave is the first rectified half-wave and the green wave is the inverted (reversed) second half-wave. Half-wave rectification (previous chapter) is important because we can use it to block one direction, but full-wave rectification also has an application. With a simple filter capacitor, we can convert this purple and green wave into a pure DC wave (see Figure 14). So here we have created an AC to DC converter, similar to a UPS (uninterruptible power supply) for charging batteries from AC power. Most of the applications are in the field of power electronics.

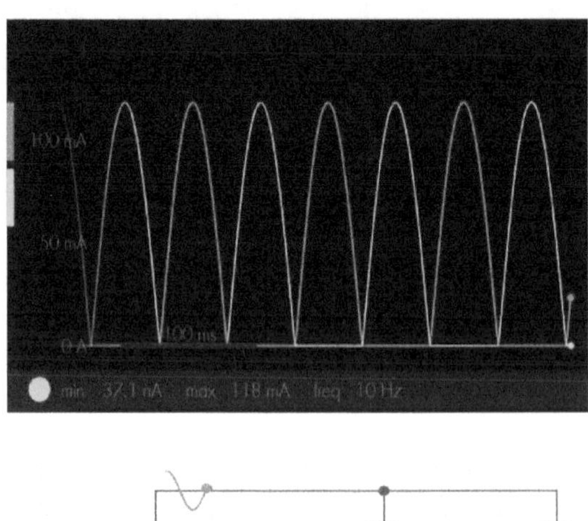

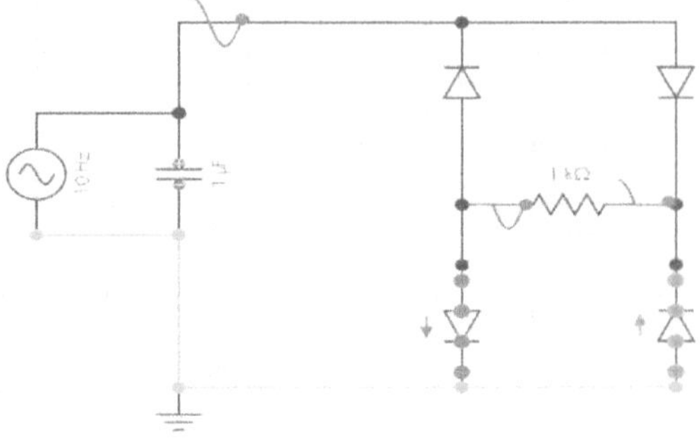

Figure 13: Full wave rectifier - two waves (violet and green) at the output side

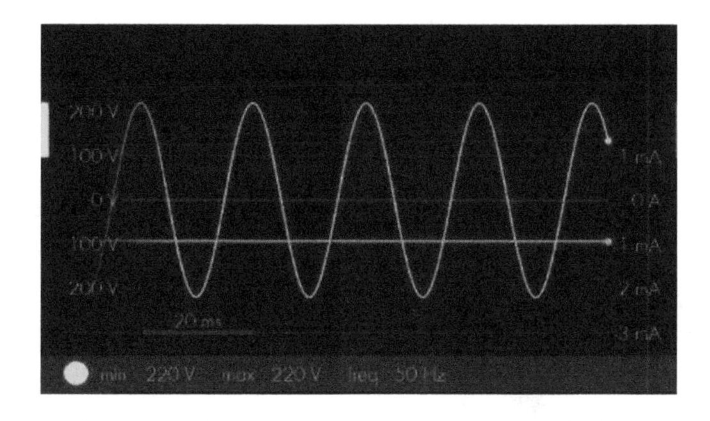

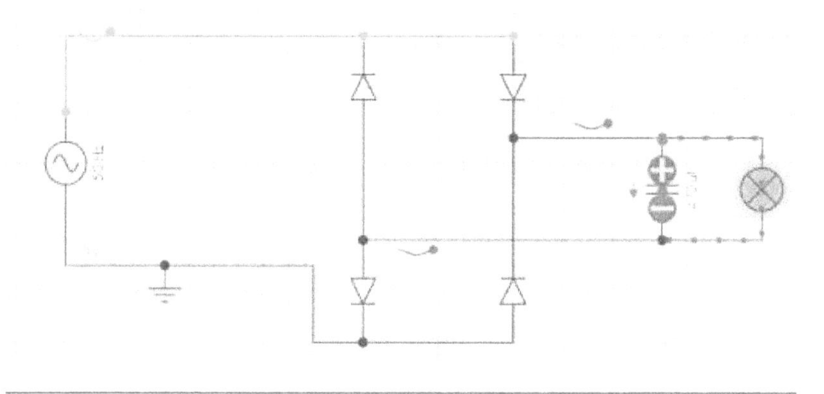

Figure 14: Bridge rectifier with filter capacitor and pure DC output (green wave)

3.3 What is a transistor?

A transistor is a simple three-terminal device that is best thought of as a valve that controls the flow of water in a pump. When we turn the control wheel of the valve in a certain direction, i.e. open, the water flow increases and when we turn it in the other direction, i.e. close, the flow decreases. The valve in the case of the transistor would be diodes and the water would be the current. Electronics in general, to put it simply, has a lot to do with switching elements and transistors also behave like a switch. In addition to this switching capability, transistors also have the property of amplification, which would be equivalent to changing the valve ratio for the amount of water output. This gain property is particularly important in the world of electronics. In analog systems, we use the operational amplifier (op-amp; dc-coupled amplifier) for this purpose, and transistors can be used to achieve an equivalent element to this in digital systems as well. There are different types of transistors, one of the simplest is the bipolar transistor (**BJT**). Furthermore, here we will discuss the field effect transistor (**FET**) and the metal oxide semiconductor field effect transistor (**MOSFET**). All types of transistors have their special characteristics and are used in different applications.

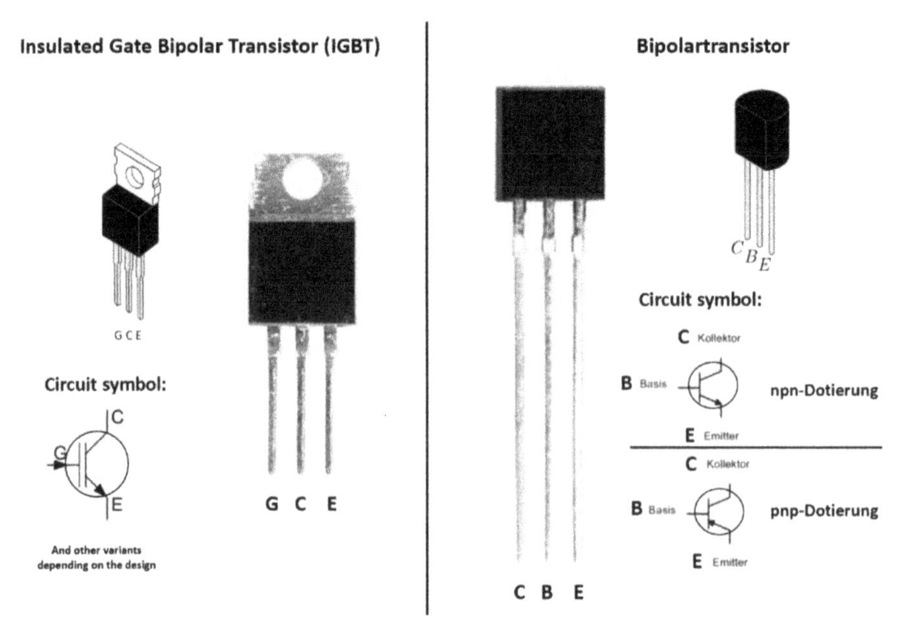

Figure 15: Two variants of transistors with their connections and circuit symbols (left: IGBT and right: bipolar transistor)

3.3.1 The Bipolar Transistor (BJT) - Basics

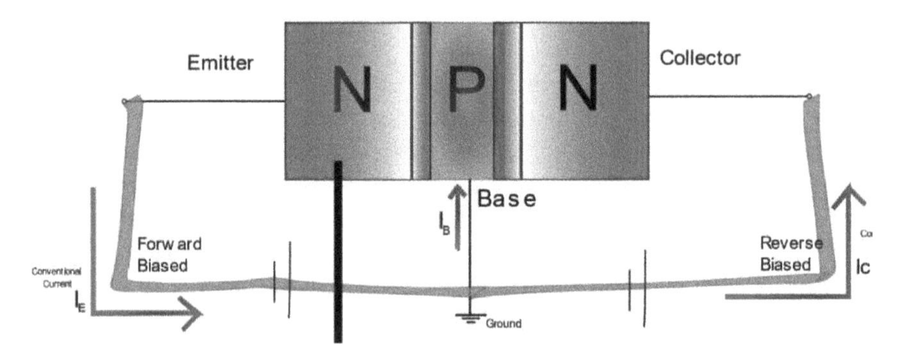

Figure 16: npn bipolar transistor in normal operation

A PN junction is a combination of doubly doped semiconductors. When we have three junctions combined in a sequential manner, we get a device with very special properties. Such a device with two combined diodes is commonly called a bipolar transistor "Bipolar Junction Transistor" (BJT). Because of the double junction it is possible to build a BJT in two ways: once with two N-doped elements **(NPN)** and once with two P-doped elements **(PNP)**. The **normal transistor action** (fundamental current flow) occurs when one of these junctions is forward biased and the other is reverse biased (Figure 16). The middle region, located between two highly doped regions, is called the **base (B)** and of the other two regions, one is called the **emitter (E)** and the other is called the **collector (C)**. Generally, the emitter is a junction where the junction is forward biased and the collector is the junction where we have reverse biased junction with the base. An important point to note here is that the sandwiched base is not highly doped (this is why in Figure 16 the base is shown smaller than the other two regions). This lightly doped base is what makes the transistor a special device. Just as with the valve, we can now control a large collector current with this base.

The theory behind BJTs is simple, especially if we know how a PN junction works. In our case, we assume an NPN-BJT transistor (Figure 16). When we connect the first NP-emitter base region in the forward direction, free electrons from N break through the junction and flow toward the P region. Since the base is only lightly doped, some of these electrons connect to holes, completing the circuit, while the remaining electrons remain in place. If we then connect another power supply to the base collector so that the junction goes into reverse voltage, the remaining electrons thereby continue their flow from the base through the collector and then flow towards this high voltage.

The current splits into base and collector starting from the emitter. If we increase the base current (i.e. more electrons connect to P-holes in a considered time period), effectively more remaining electrons flow to the collector (in a considered time period). Remember these two simple statements - to solve problems we can apply them well. Below are a few important mathematical relationships for BIPs:

$$I_E + I_B = I_C \tag{2-1}$$

α indicates the quality of the transistor

$$\alpha = \frac{I_C}{I_E} \tag{2-2}$$

β is the gain (amplification factor)

$$\beta = \frac{I_C}{I_B} \tag{2-3}$$

A high value α of a transistor indicates that relatively less base current flows, since in this case the collector current approaches the emitter current. In practice, the value of α is between 0.95-0.99 (low-power signal transistor) and ideally it should be 1. In the same way, the value β usually has a value of 100-150. This β represents the gain of BJTs and is therefore also called gain factor.

Generally, BJTs are used in three **configurations** called common base **(CB)**, common emitter **(CE)**, and common collector **(CC)** configurations. Common is related to reference or ground. For a gain (β), we use the CE configuration because the output voltage (U_{CE}) comes from the gain of the input (U_{BE}). In the same way, other configurations have some special use cases. To control the level of bias voltage (corresponds to the output parameters of the circuit), we use circuits usually called **bias circuits.** Here we use some resistors to control the level of bias voltage. To design circuits in which transistors are present, the values of these resistors must be set to the desired parameters.

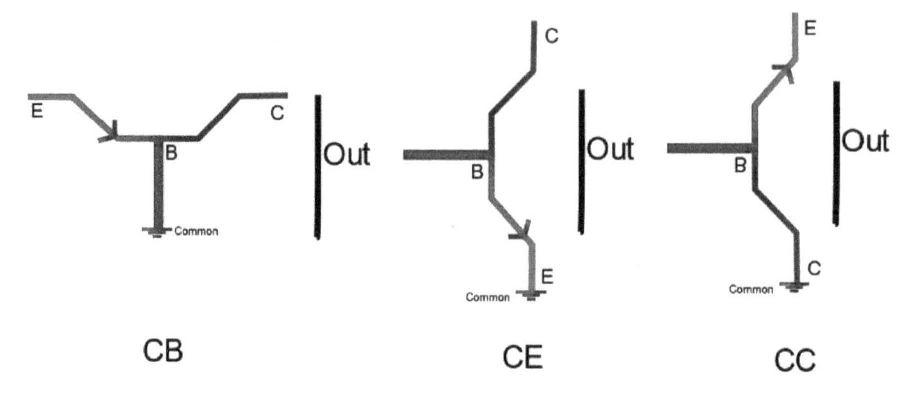

Figure 17: BJT configurations

3.3.2 The Junction Field Effect Transistor (JFET) - Basics

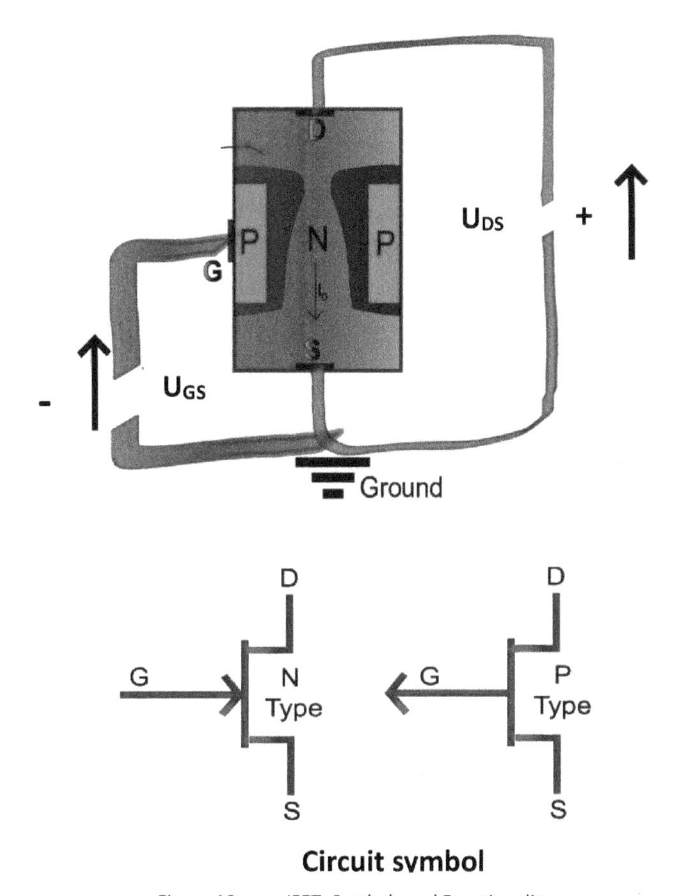

Figure 18: npn-JFET: Symbols and Functionality

The **junction field effect transistor (JFET)** is also a three-terminal transistor device whose current is controlled by the third terminal, called the gate. In this context, the other two terminals are called **source** (inflow) and **drain** (outflow). Just like BJTs, a JFET can be thought of in two ways: once with an N-channel sandwiched between two P-channels, and once in the opposite way (P-channel sandwiched between two N-channels). But unlike BJTs where current flow is through both electrons and holes, in N-channel JFET current flow is through electrons and in P-channel JFET current flow is through holes. Let's take a closer look at this in the N-channel case:

CASE I : U_{GS}= 0: The electrons in the N-channel, which flow in from the **source,** move in the direction of the **drain** (see Figure 18). Drain is here connected to the positive pole of the battery with source ground. Now, when these electrons of N channel flow towards drain (+ terminal), some of them try to escape through P region and thus a junction (depletion zone) is formed between N and P. The width of this junction depends on the drain-source voltage (**U_{DS}**). As we increase this voltage U_{DS} , the free carriers in the junction increase (as in BJTs) and thus block other carriers with more potential, resulting in a larger width. We can increase this voltage U_{DS} up to a certain point, which is called **pinch-off voltage** or **pinch-off voltage U_P, which** corresponds to an equilibrium point. If we increase this voltage further, the electron flow in the N channel towards the drain will stop increasing, and thus there will be no effect on the drain-source current **I_{DS}**. This constant current after the pinch-off voltage is reached in this saturation region is called **drain saturation current I_{DSS}** (drain-to-source current with the gate shorted).

Now why do we have an undistributed junction length in Figure 18? Because the drain voltage in the N-channel cannot be constant everywhere when current is flowing through it. For a given voltage U_{DS} (say 5 V), this voltage U_{DS} decreases with conventional current flow from drain to source along the path, effectively reducing the junction width. So in JFETs, we have a high width at the drain side, which gradually decreases towards the source.

Case II : $U_{GS} \neq 0$: If we now connect the gate to the negative terminal, it attracts the holes in an already widened junction. This effectively reduces the width of this junction. If we then increase the voltage U_{GS} more current flows from the drain to the source (i.e. I_{DS} increases). Thus, in this case, the pinch-off voltage U_P also increases and so does the drain saturation current **I_{DSS}**.

These two statements in the last two paragraphs can be represented by the following equation. We can also represent these properties in a diagram (see Figure 19; "V" stands for the voltage "U" in this figure).

$$I_D = I_{DSS} \left(1 - \frac{U_{GS}}{U_P}\right) \qquad \text{2-4}$$

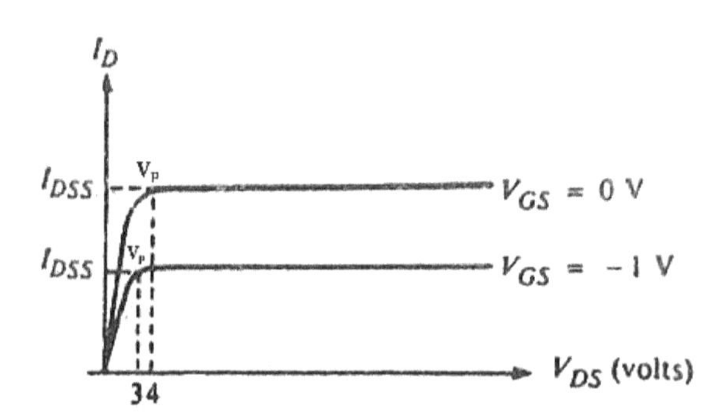

Figure 19: Quadratic law characteristic curve

3.3.3 The Metal Oxide Semiconductor Field Effect Transistor (MOS-FET)

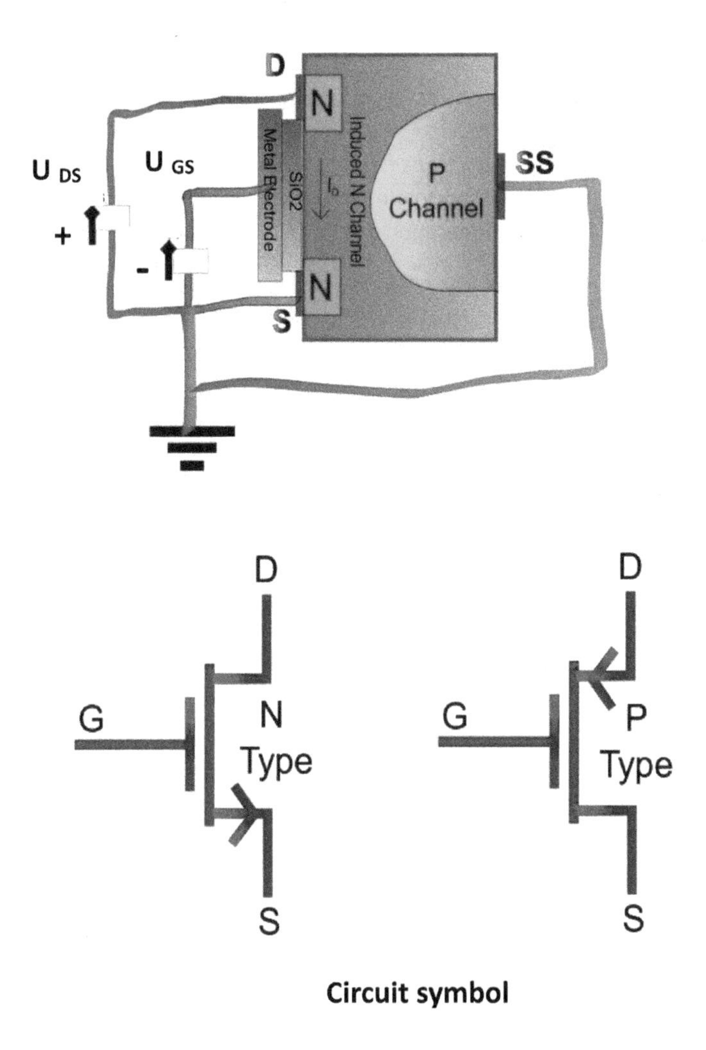

Figure 20: Enhancement-type N-channel MOS-FET (SS is substrate)

Metal oxide semiconductor field effect transistors (MOSFETs) are like a successor or updated device compared to the previously mentioned transistor types. MOSFETs are the most common type of transistor and are used in many applications, especially those that require high power. A BJT is a current controlled device because the collector current is controlled by the very low base current. Its applications are therefore in low power electronics. However, MOSFETs have now also found their use in low power electronics as they have a high switching capability. In terms of applications, it can generally be said that MOSFETs are better and more versatile than BJTs.

In general, MOSFETs are comparable to JFETs, there is only a small difference in the metal oxide insulation. The structure is also somewhat different from JFETs. However, we will not go into the theoretical details of MOSFETs here because the operating principle of this device is pretty much the same as JFETs. In N-channel MOSFETs, there is a P-substrate from which the electrons flow to the isolated N-channel. As mentioned earlier, the electrons in the isolated channel are induced by the P-channel substrate. This is important for the normal operation of MOSFETs.

In MOSFETs, the insulated gate layer separates the gate, so unlike JFETs, the bias voltage does not matter in this case. In JFETs, the gate must be reverse biased to prevent current from flowing through it. This helps the JFET to control the drain current with the gate voltage. Because of the metal oxide insulation, MOSFETs operate with zero bias voltage. There are two types of MOSFETs: a **depletion type** and an **enhancement type**. Both types can be doped in two ways (with N and P channels). The enhancement **type** (normally open) requires a control voltage U_{GS} to turn the device on, while the **depletion type** (normally closed) requires the opposite (turn it off). Enhancement type MOSFETs are the most commonly used. It is also important to note that JFETs can only operate in depletion mode, where a negative voltage is applied to the gate. So if you increase the voltage on this one, the device will turn "OFF". MOSFETs, on the other hand, can operate in both modes as described. MOSFETs also have the same characteristics in terms of the characteristic curve, as shown in Figure 19. For MOSFETs, instead of pinch-off voltage / pinch-off voltage, we use the term threshold voltage U_T.

Figure 21: MOSFETs can look like this figure, for example

3.4 Practical applications of electrical engineering and electronics

The general introduction of individual electronic components has been the main goal of this chapter so far, so we haven't had any circuit problems in this chapter yet. In general, circuit calculations here also work with the basic circuit rules we learned in previous chapters. Instead of solving theoretical computational problems, in the following we will look at some basics that we can use to design circuits.

Sample Example 4

Uninterruptible Power Supply - Design a circuit that switches to battery power when power is turned off, and back to power when it is turned back on. Goal: We want to have uninterrupted 12 V at the output to power a laptop charger that requires this voltage. The 12 V battery is to serve as a kind of emergency power supply.

Notes:

1. A **relay** is an electrical component generally used to switch things electrically. When we give a signal, the electromagnetic switch changes its position to another terminal and switches things accordingly. When there is no electrical signal, the closed position is called normally closed (NC) and the open position is called normally open (NO).
2. A light emitting diode (**LED**), as we already know, is a diode and therefore allows current to flow in one direction. When current flows through the LED semiconductor diode, electrons combine with holes and release energy in the form of light.
3. A **capacitor** is a component that can store electrical charge (and the energy). More details about this in the next chapter.

Design Procedure:

First we need a relay that switches to the NC position when the power is turned on. Then we connect a battery to this NO position and a laptop charger to its output.

Finally, we add two reverse-bias diodes to prevent battery current from flowing back to the source.

The top figure on the next page we see how the LED is powered by the main power supply and on the bottom figure how the LED is powered by the battery. The main power supply is 13.5 V. Finding a battery with this voltage is not possible in practice, so we can use an amplifier (such as BJT amplifier) or a boost converter. In fact, we used a capacitor to store the charge so that when the relay is turned off, the load remains on the supply. 1μF is quite a low value in this case, so in practice you can use a capacitor of 1mF. What exactly a capacitor is and how it works will be explained in detail in the next chapter.

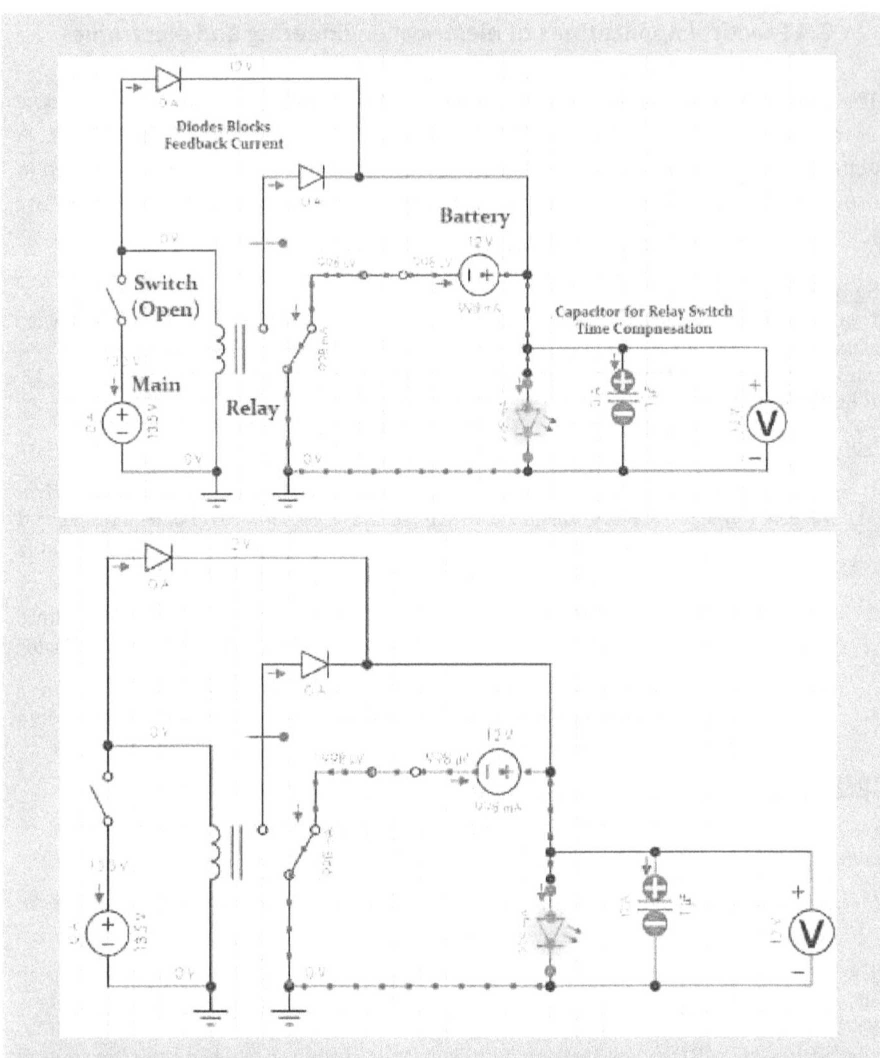

Before the development of MOSFETs and other semiconductor devices, there was only relay(computer). But now, with this efficient integrated technology, we can use semiconductors for switching. Therefore, the above circuit can also be drawn with a MOSFET and a BJT. While the main current switches the BJT, it controls the output current through its base. The MOSFET remains off in this case. When we turn off this main switch, the battery triggers the MOSFET gate, and the load is handled by the battery. For the 13.5 V voltage, we can again use either an amplifier or a boost converter. So the above circuit can also be designed with solid state devices instead of relays as shown in the figures on the next page.

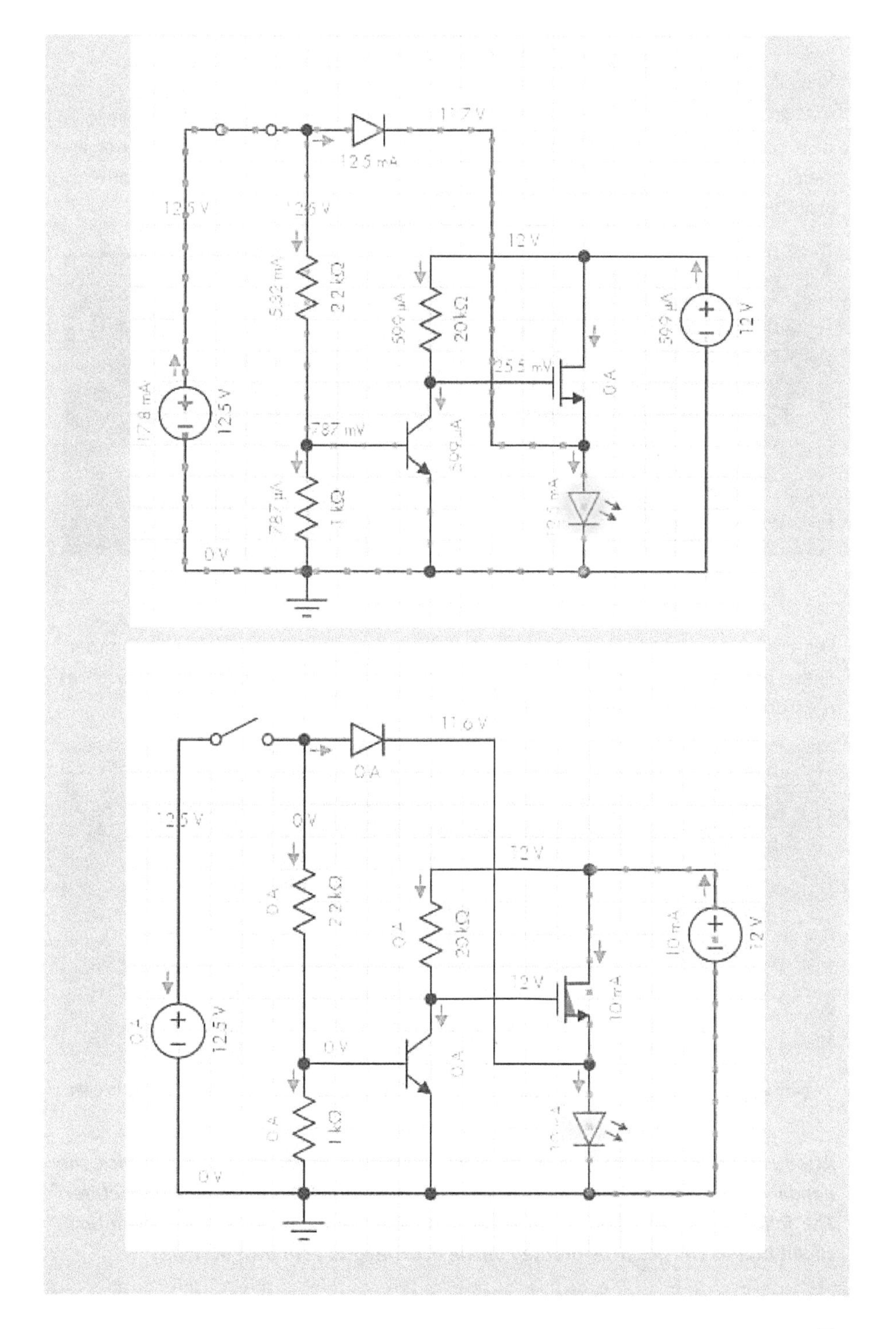
12.5 mA
532 mA
22 kΩ
599 µA
20kΩ
12 V
787 mV
787 µA
1 kΩ
0 V
12.5 V
0 A
10 mA

3.4.1 Printed circuit board (PCB)

We can design such circuits in reality in a practical way on a breadboard, also called a stripboard, for prototyping. Such stripboards are not permanent solutions, but are mainly used to easily attach different electronic components by hand in terms of prototyping.

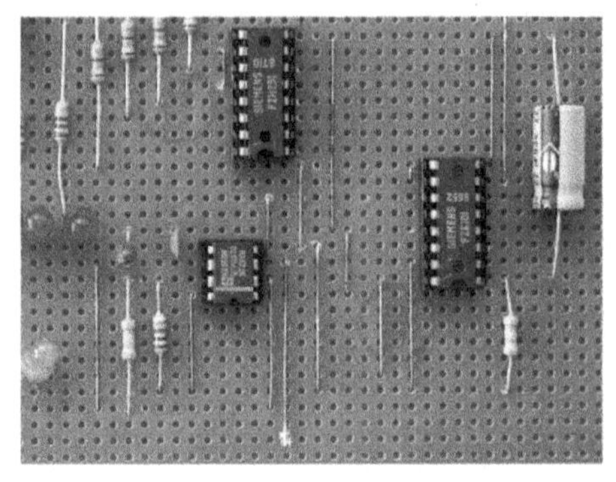

Figure 22: Prototyping a circuit with a stripboard

For a reliable and durable solution, printed circuit boards (PCB) are the most common. These boards simply serve as a carrier element for electronic components such as resistors, transistors, capacitors, etc.

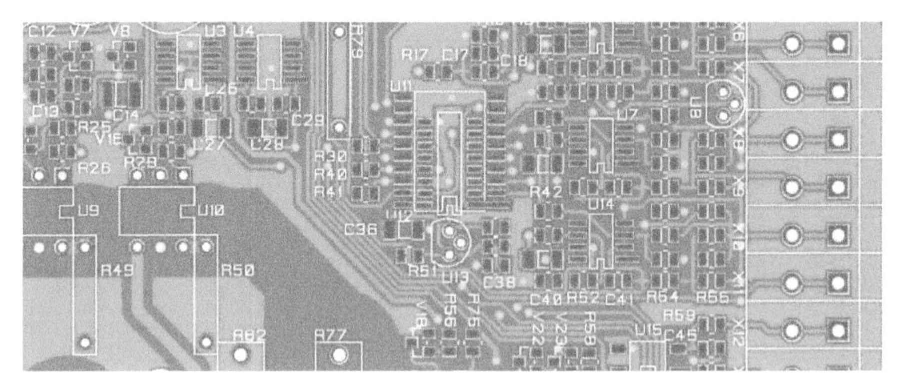

Figure 23: A printed circuit board / circuit board (meanwhile also online orderable according to own wishes)

Attention: Please do not try to produce printed circuit boards without experience, the handling of etchants is dangerous, follow the safety regulations of the manufacturer! The following information is not a recommendation for action! In the meantime, circuit boards can easily be ordered online according to your own designs.

To design a circuit board, for example, we can use computer programs to draw circuits virtually on circuit boards and then print them mirrored on glossy paper. The glossy paper then transfers the circuits of this print to the circuit board when it is heated. Diluted FeCl2 solution is then used to remove the insulation from printed lines. The removal of insulation from conductive wires is commonly referred to as **etching.** The **etching process** removes excess copper on the board, leaving the desired conductive traces. After cleaning and drying, holes can be drilled, the necessary components such as resistors, transistors, etc. added, and you then have a working circuit board. Of course, in industry, such circuit boards are not made manually but by machine. Cell phones, computers, TV remotes, and all other electronic devices have circuit boards as the essential substrate for components to interact in circuits and perform the functions we want.

3.4.2 The multimeter: Current and voltage measurement in practice

In electrical engineering practice, we often use multimeters as measuring instruments. Multimeters with two terminals can measure voltage, current, resistance, capacitance and inductance. They can also measure the polarity of transistors and perform a continuity test with them. The continuity test tells us if a circuit is shorted or not. Multimeters can only measure one variable at a time (like current or voltage). To measure multiple parameters, we need to use several individual devices. In the figure below, a simple multimeter is shown with the different measurement ranges. Depending on what you want to measure, you turn the dial to the appropriate range.

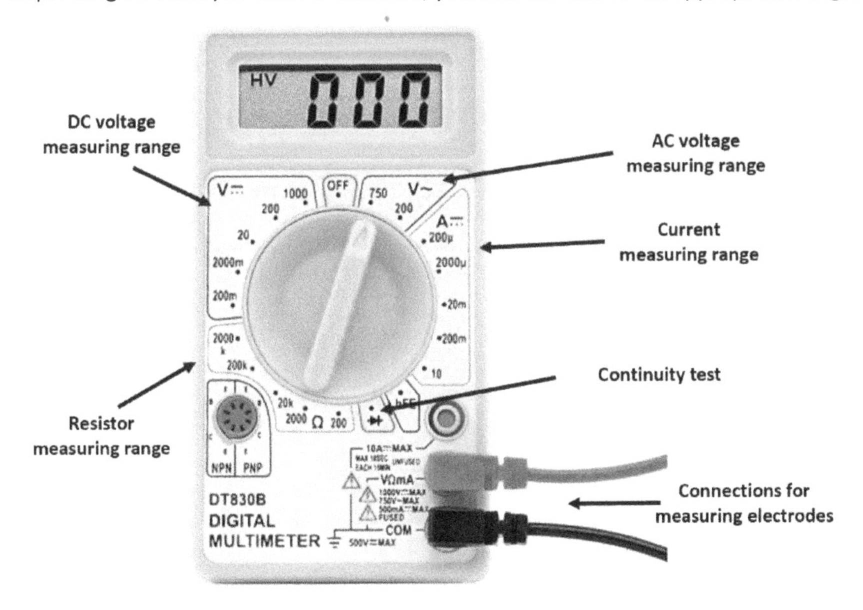

When taking a measurement, always start from the highest possible voltage or amperage or resistance value and then turn the display setting down until a suitable value is displayed. This means, for example, that if you make a measurement on a DC voltage source and you suspect a value between 20 and 200 V, you turn the setting range to 200 volts.

If you want to measure a voltage, you have to connect the measuring electrodes parallel to the voltage source or to the component you want to measure. In the case of a light bulb, for example, this would work like this:

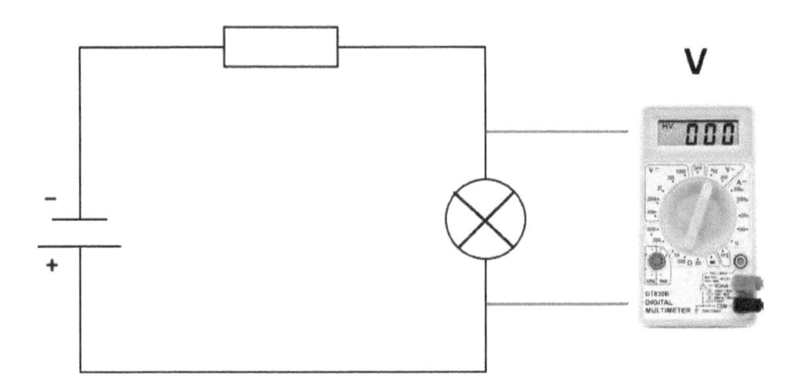

And if you want to measure the amperage of a consumer, you have to connect the measuring instrument (multimeter) in series to the consumer, i.e. disconnect the line. This would work like in the following:

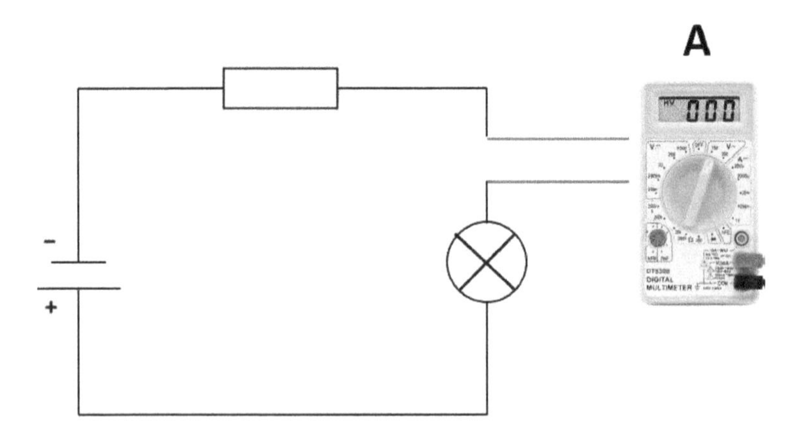

4 Direct current vs. alternating current & sine waves

4.1 Introduction to the topic

What is current and what is voltage? What is direct current and what is alternating current? These generally known terms are often used without knowing exactly what is behind them. So far, we have only dealt with direct current. In this chapter we will take a closer look at the basic differences. Current can simply be thought of as the flow of charges in an electrical conductor. Voltage, on the other hand, is simply the energy of a charge flowing in that conductor. Alternating current, as the name suggests, is a type of current that changes its direction of flow (polarity) over time. This happens periodically, i.e. recurrently. This is not the case with direct current, where the direction of flow (polarity) remains the same over time. This applies equally to the terms AC voltage and DC voltage, although here it is the voltage that is being considered.

So far we have studied circuits that work with direct current. For example, we looked at resistors in the first chapter. However, there are also elements in electronics that require a change in voltage and current for normal operation. These elements (**capacitors** and **inductors**) also have some important applications in electrical and telecommunications engineering. For a good understanding of how they work, we will first deal with sine waves in the following, since these periodic waves represent the continuous change in voltage or current and form the basis for their description (see Figure 24). **Alternating current** (sine waves) is used in the transmission (e.g. through a power pole) of electric current. Sine waves are also of great importance in the study of information and communication and offer very interesting properties. **Direct current** is nowadays mostly restricted to low voltage applications. Direct current cannot encode information because it flows in only one direction. Why direct current is not efficient for the purpose of transmitting power (e.g., power pole) is also an interesting topic that we will discuss in the next chapter. This chapter, as mentioned earlier, is first devoted to the study of sine waves. We will also cover **capacitors, inductors** (inductors) and **RLC circuits in this chapter to** match.

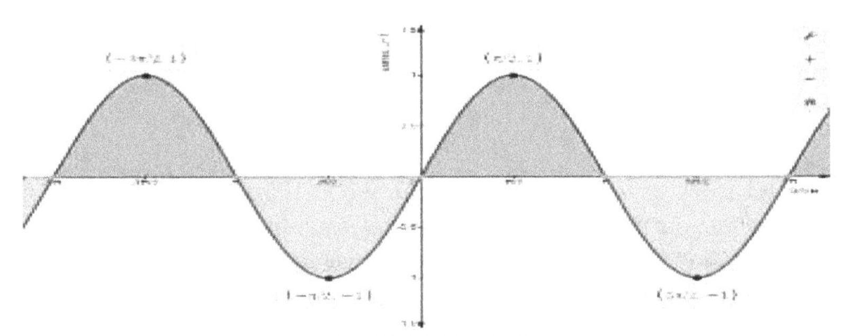

Figure 24: A sine wave

4.2 Alternating voltage / alternating current

Alternating voltage / alternating current (AC) varies sinusoidally with time. Figure 25 shows a sine wave. These types of waves have a **time period** and a **so-called phase**. The period, simply put, is the time after which the pattern of a wave repeats. A simple sine function $f(x) = sin(x)$ has a period of 2π. The term phase, simply put, describes the displacement of a wave. For example, the black wave in Figure 25 is shifted to the right on the x-axis by $\frac{\pi}{2}$ to the right on the x-axis, so its phase is $\frac{\pi}{2}$. In the formula of the function, this is expressed by the term $-\frac{\pi}{2}$ term.

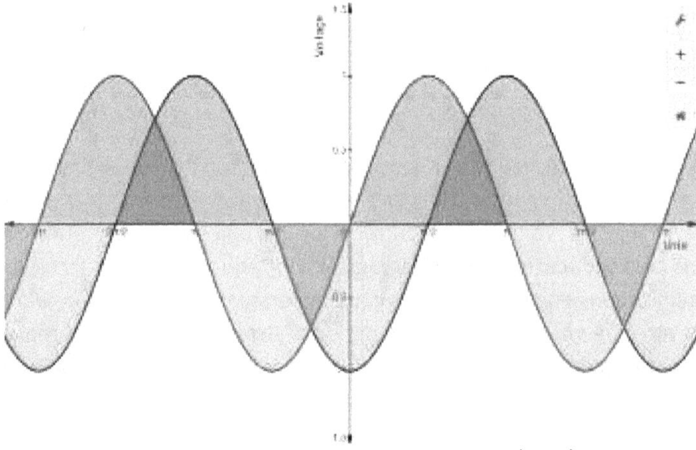

Figure 25: Representation of $\sin(x)$ *in purple and* $\sin\left(x - \frac{\pi}{2}\right)$ *in black*

The **frequency of** a sine wave is the reciprocal of its period. In general, the frequency indicates the degree of compression or expansion of a wave. You can also say that frequency indicates the number of periods in a second. Frequency is measured in Hertz, named after the German physicist Heinrich Hertz. 1 Hz corresponds to one oscillation per second, i.e. 1/s. For example, the black wave in Figure 26 is now more stretched than the original wave (purple), and its frequency is only half the frequency of the original purple wave. The frequency of household electricity is 50 Hz or 60 Hz and varies by region. 50 Hz means that a cycle can repeat 50 times in one second. 50 Hz also means that the AC wave crosses the zero-sequence voltage (x-axis in the coordinate system) 50 times in one second, as the direction of the AC current/voltage changes as already known. This basically means that a light bulb turns off 50 times in one second. However, these fluctuations are not perceptible to the human eye as they are too fast. However, we could make these fluctuations visible with a slow motion camera.

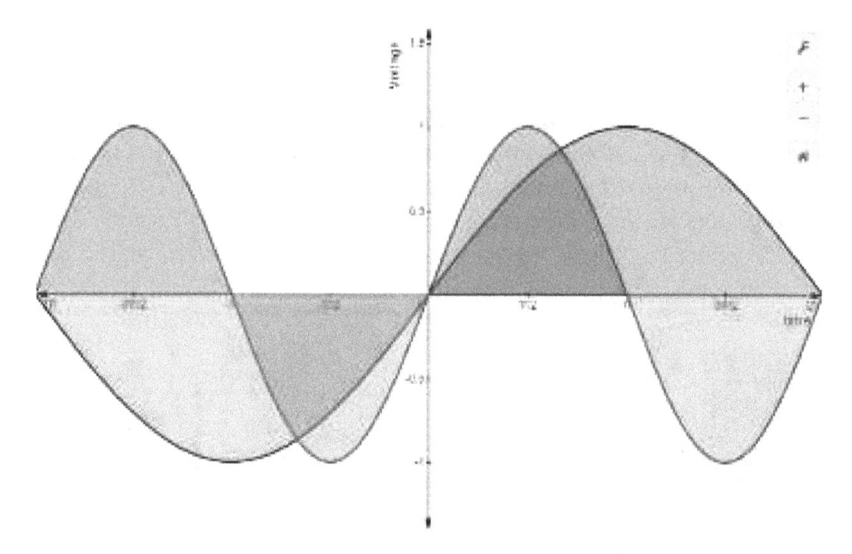

Figure 26: Representation of $\mathbf{sin}(\boldsymbol{x})$ *in purple and* $\mathbf{sin}\left(\frac{x}{2}\right)$ *in black*

You should remember the following important relationships in connection with waves:

1) The frequency of $sin(2 \cdot x)$ is twice as high as $sin(x)$, the period is half as large.

2) The period of $sin\left(x - \frac{\pi}{2}\right)$ is delayed (shifted) by $\frac{\pi}{2}$ delayed (shifted) than as in the original $sin(x)$

4.3 Series resonant circuit (RLC circuit)

In addition to resistors, there are several other elements that are fundamental to electronic circuits, namely **capacitors** and **inductors** or **inductors**. **A circuit with a combination of resistors, capacitors and inductors is generally referred to as a series resonant circuit or RLC circuit.** "R" stands for resistor, "L" for coil (inductor), and "C" for capacitor. The study of RLC circuits is a complex subject and requires a separate chapter. In this section, we will try to cover some basics about this circuit and its associated components.

4.3.1 Capacitors

In very simple terms, a capacitor consists of nothing more than two plates arranged parallel to each other and a dielectric between them. A dielectric is simply a weakly or non-conductive substance (solid, liquid, gas) with charge carriers that are not free to move. Capacitors are generally considered charge storage devices because, when an electrical potential is applied, they can store voltage (energy) in their plates.

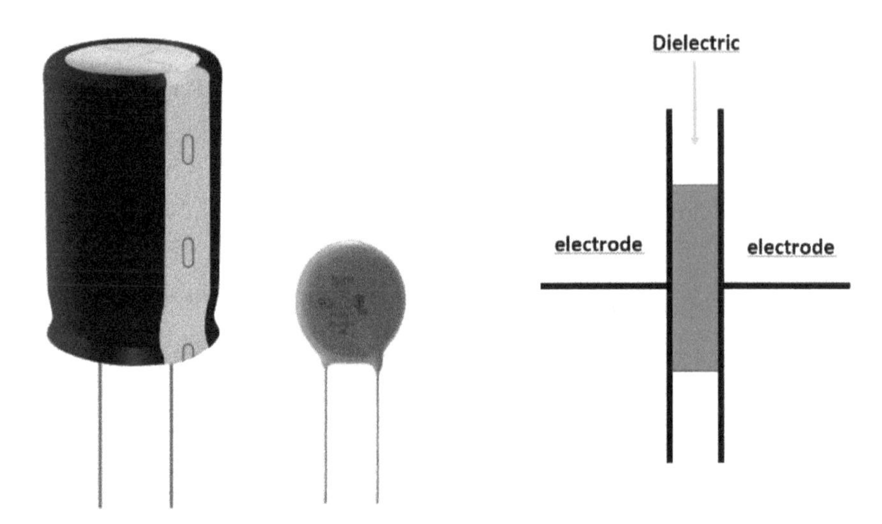

Figure 27: Two types of capacitors: electrolytic capacitor (left) and single-layer ceramic capacitor (middle); and the general schematic structure of a capacitor (right)

The ability of a capacitor to store charge depends on the area (A) of the plates, their spacing (d), and the dielectric (ε) between the plates. This ability to store charge is commonly referred to as **capacitance.** $\left(C = \frac{A}{\varepsilon d}\right)$, "ε" here is called the **dielectric constant**. As the charge in the plates increases, the voltage of the capacitor also increases, and this continues until the capacitance is reached. This statement is described mathematically by equation 4-2:

$$Q(t) \propto U_c(t) \Rightarrow Q(t) = C \cdot U_c(t) \qquad \text{4-2}$$

In terms of voltage and current, we can rewrite this equation to:

$$\frac{dQ(t)}{dt} = C \cdot \frac{dU_c(t)}{dt} \Rightarrow I_c(t) = C \cdot \frac{dU_c(t)}{dt} \qquad \text{4-3}$$

As a reminder, current is the flow of charges per unit time.

Equation 4-3 is interesting because it shows that current only flows in a capacitor when the voltage changes. At constant voltage, the capacitor behaves like an open circuit in which, of course, no current can flow.

Situation - DC current: If we connect a capacitor in series to a DC voltage source, the charging of the plates begins. In the course of charging, the capacitor repels the source electron and thus stops the current flow.

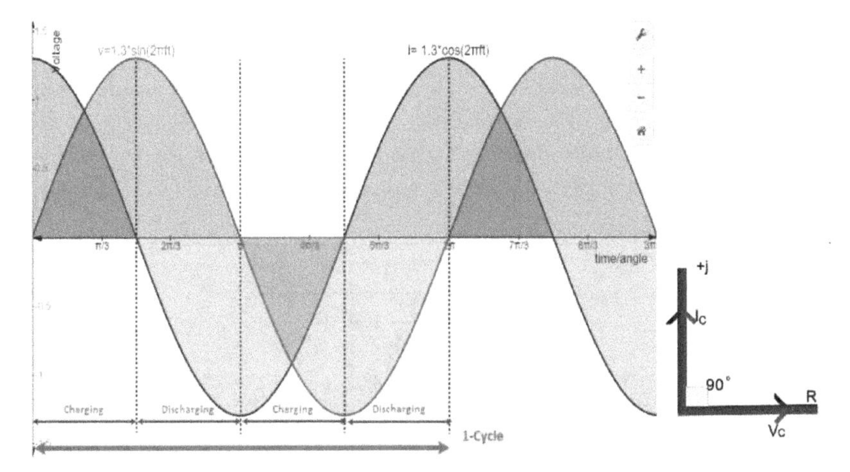

Figure 28: : Voltage wave (sine) and current wave (cos) of the capacitor (current wave "lags" the voltage by $\frac{\pi}{2}$ behind the voltage. The phase diagram of the capacitor is shown on the far right.

Situation - alternating current: If we connect this capacitor to an alternating current source, in the first quarter cycle (from 0 to $\frac{\pi}{2}$; see Figure 28) it takes up voltage (i.e. stores charge), which it then releases in the second quarter. In the negative half cycle of the wave, this action is repeated, but now with opposite polarity, i.e. the capacitor charges in the first quarter and then discharges in the second quarter.

It is important to note that the capacitor "absorbs" voltage, but at the same time has no influence on the current. It is this absorption that causes a delay in the voltage, and therefore the voltage in the capacitor "lags" the current by $\frac{\pi}{2}$ behind the current. The right side of the above figure shows the phase diagram of the capacitor. As we will see, phase diagrams are useful in calculating the voltage and current relationship of capacitors.

Capacitive reactance:

Assuming we increase the frequency of the AC source, more electrons will flow through the capacitor in a unit of time, resulting in an increase in current. This increase in current decreases the resistance of the capacitor. In the same way, the resistance of the capacitor increases when the frequency decreases. This so-called reactance of a capacitor is commonly referred to as (capacitive) **reactance** (X_c), and Equation 4-6 relates it to the AC frequency as follows:

$$X_c = \frac{1}{2\pi f C} \tag{4-6}$$

Series and parallel connection of capacitors:

From the previous explanation and the formula above, we can see that the capacitance behaves inversely (i.e. inversely) to its reactance. Therefore, in the parallel circuit we add the capacitance and in the series circuit we add the reciprocal of the sum of the reciprocals of the capacitances, i.e. exactly opposite to the series and parallel connection of resistors.

Series connection:

$$C_{eq} = \frac{1}{(\frac{1}{C_{eq_1}} + \frac{1}{C_{eq_2}} + \cdots + \frac{1}{C_{eq_n}})} = (\frac{1}{C_{eq_1}} + \frac{1}{C_{eq_2}} + \cdots + \frac{1}{C_{eq_n}})^{-1} \tag{4-7}$$

Parallel connection:

$$C_{eq} = C_1 + C_2 + \cdots + C_n \tag{4-8}$$

4.3.2 Inductors (coils)

Every current-carrying, conducting wire generates a magnetic field around it. So what happens when we wind a whole coil from a conducting wire? Let's take a closer look at that in this chapter. Such a wire wound into a coil is called an inductor.

Figure 29: Toroid coil with iron core (left); coil (right)

So what special thing can happen when we wind a wire or other conductor into a coil? In short, it changes the magnetic field produced. When current flows through this wound wire, it creates a change in the magnetic field and when that current changes, the magnetic field created opposes the new current so it can't change any further. As an analogy, you can think of an inductor as a mill wheel with water flowing through it with some potential energy. Initially, the (static) friction of the wheel opposes the energy of the water, but as soon as the water drives the wheel (breakaway torque) and the static friction changes to sliding friction and is thus less than the driving force, the wheel "assists" the movement of the water in a certain way due to the inertia of the wheel. If now suddenly no more water flows, the movement continues (inertia) until the friction becomes too "high" for the remaining energy.

What is such a component? An inductor is a component that can be considered the counterpart of the capacitor and whose operation can be explained by **Lenz's rule.** Lenz's rule states that the current generated by induction is always directed against the causer (magnetic field). A mechanical force effect that may arise is called: **Lorentz force.**

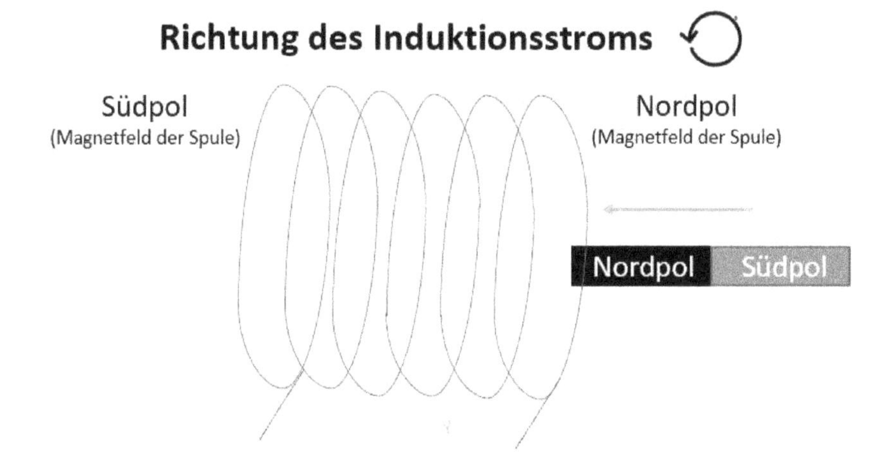

Figure 30: Lenz's rule applied in practice. If a magnet is moved towards the coil as shown, the coil will move to the left, because the induction current in the coil is directed in such a way that a north pole is created on the right (same poles repel each other). Exactly the opposite is also true for the other case (north pole and south pole; attraction instead of repulsion).

Just as with capacitors, inductors have a constant, the inductance, which tells us the ability of an inductor to store energy in the magnetic field. This depends on the number of turns and the dimensions, such as length, of the coil ($L = k \cdot N^2$) "k" here is simply a constant for the dimensions. "N" stands for the number of turns.

Just as with capacitors, the magnetic flux of the inductor increases with increasing current in the coil until the inductance is reached. In mathematical terms, this means:

$$\lambda(t) = L \cdot I_L(t) \quad 4\text{-}9$$

And a simple derivation by time on both sides gives:

$$\frac{\lambda(t)}{dt} = L \cdot \frac{dI_L(t)}{dt} \Rightarrow U_L(t) = L \cdot \frac{dI_L(t)}{dt} \quad 4\text{-}10$$

Properties of inducers

As with capacitors, a **direct current flow** short-circuits the inductor (ideally no resistance). Since in the case of direct current there is no change in the direction of the current and therefore no change in the magnetic field, the circuit is simply short-circuited without any voltage being induced. And in the case of an **alternating current,** during the first half period of the sine wave, the inductor absorbs all the current but does not change the voltage. And because of this absorption, the current in the case of the inductance "lags" the voltage by $\frac{\pi}{2}$ behind the voltage. If you think about the capacitor again for a moment, you may notice that in the case of the capacitor, it was the other way around, that is, in the case of the capacitor, the voltage "lags" the current by this factor. Very good attention!

The reactance of the inductance X_L (also called **reactance** or more completely: **inductive reactance**) is also related to the frequency of the alternating current (equation 4-11). To do this, consider how an inductor reacts when the current flow increases or decreases in one second. We can easily derive this relationship:

inductance resistance

$$X_L = 2\pi f L \quad 4\text{-}11$$

By the way, the term reactance, generally speaking, simply means a complex resistance. However, in the case of capacitors and inductors, the phases of current and voltage differ, which in turn affects their resistances. To distinguish the resistance of these components from the resistance of a simple resistor (here the component is meant), the term capacitive or inductive reactance is therefore used.

For the circuit analysis the following laws for the series and parallel connection of inductors are again important: *(Note: Compare the following equations with those of the capacitor, then you will notice something and it will be easier for you to remember it)*

Series connection

$$L_{eq} = L_1 + L_2 + \cdots + L_n \quad 4\text{-}12$$

Parallel connection

$$L_{eq} = \frac{1}{(\frac{1}{L_{eq_1}} + \frac{1}{L_{eq_2}} + \cdots + \frac{1}{L_{eq_n}})} = (\frac{1}{L_{eq_1}} + \frac{1}{L_{eq_2}} + \cdots + \frac{1}{L_{eq_n}})^{-1} \quad 4\text{-}13$$

Series resonant circuits (RLCs), as we already know by now, are simply a combination of a resistor, a capacitor and an inductor. This allows a total of 8 scenarios ($\sum_{n=0}^{3}\binom{3}{n} = 8$) of RLC circuits, two of which are simple series and parallel circuits. These two circuits are also the most important. RLC circuits are used in electronic filters, which we in turn use for tuning radio and television channels, oscillator circuits and in audio control. Most often, RLC circuits are applied when signal analysis is required. Again, most applications of signal analysis are found in communications. **Phasors are** needed for the calculation of RLCs. Phasors are, in simple terms, the vectors for 2-dimensional complex numbers, which in this case are needed to solve for capacitors and inductors. When we connect a linear resistor in series with a capacitor (RC circuit), the voltage of the capacitor is delayed because the capacitor stores voltage. For the same voltage of "R" (resistor) and "C" (capacitor), we get a right triangle which we can decompose into its components (perpendicular and base). The same is true for the inductance. The voltage of the capacitor rushes behind and that of the inductor rushes ahead.

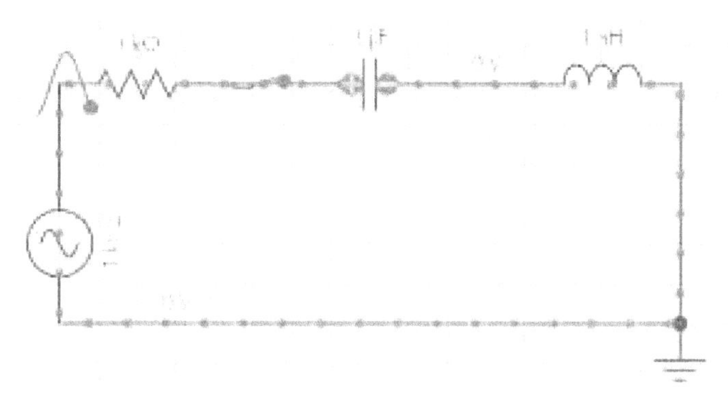

Figure 31: Series RLC circuit

As an example, consider a series RLC circuit (Figure 31). To solve for the equivalent voltage, we use the values of U_c and U_R as:

Pythagorean Theorem:

$$U_s = \sqrt{U_R^2 + (U_L - U_C)^2} \quad (4\text{-}14)$$

And for the angle in between, we can use the traditional triangle formula as follows:

$$tan(\theta) = \frac{Gegenkathete}{Ankathete}$$

$$\theta = tan^{-1}\frac{U_L - U_C}{U_R} \quad (4\text{-}15)$$

It is also important to redefine resistance here, as we are looking at a combination of resistance (actual resistive component) and reactance (capacitor & inductor). This "total resistance" in RLC circuits is called **impedance Z** and includes real and complex quantities. In simpler terms, we can also say that resistance and reactance are special cases of impedance. A resistance (R = Resistance) is described with real quantities and a reactance with complex quantities.

Formulas 4 -14and 4-15 apply to RLCs connected in series (series RLC).

5 Power supply systems

In this chapter we would like to deal mainly with the transmission of current by alternating current. Direct current could also be used for power transmission and this was also the case a long time ago. But due to the high losses over long distances when using direct current, something else had to be thought of, because such a power transmission was not possible on a large scale. It had long been known that direct current could not be efficient over long distances, and so the concept of alternating current distribution was introduced. Because of the high voltage used for this purpose, the dangers of alternating current were a major hurdle in its introduction. Nevertheless, alternating current provides cheap electricity, whereas direct current would be inefficient. While the AC distribution system itself is more complex and costly than DC would be, as it requires transformers at each end of the chain, in the end, AC transmission still pays for itself in many ways.

5.1 Energy and system of units

The electric voltage is the energy of transported charged particles. The term power $P = U \cdot I$ is used to refer to the "energy" or "work" of charged particles flowing through a conductor. Usually the term energy is defined here as the power consumed over a period of time.

There are two opposing factors in power systems. These are load demand and generation, since for maximum efficiency the generators must operate at their rated values. We have already learned about kWh (energy) calculation in the first chapter. A small example to remember: a 100 W ceiling fan running eight hours a day for a month (30 days) will consume: $100\ W \cdot (8\ h \cdot 30) = 24.000\ Wh = 24\ kWh$. The price of one unit (1 kWh) for end users is determined by the electricity tariff of the electricity supplier. Currently, this averages around 30 cents per kWh. Depending on the utility company, there is a different tariff. The tariff can be different for different consumers (e.g. commercial or residential), but it can also be based on the power factor, maximum demand or consumption per unit. Most often, it is based on consumption per unit, meaning that the more units consumed, the lower the price per unit. In some countries, where electricity demand is higher than electricity generation, the opposite is true and tariffs become more expensive as consumption increases.

In order to track the load changes in a unit of time, so-called **load curves** are used (mostly in distribution networks). Distribution stations use these load curves to get an idea of the monthly demand and power plants use them for optimal generation. For low cost "generation" of electricity, care is taken to keep load demand and generation at similar levels. Indeed, a major weakness with respect to electricity is that we cannot store electrical energy directly (except in battery storage and with the help of conversion to other forms of energy, such as hydroelectric potential energy). In practice, especially in 3-phase systems, which we will learn about later in this chapter, it can be difficult to manage all generators with dynamic consumer data and unpredictable consumption trends. In renewable energy systems where the flow of electricity is bidirectional (i.e. from consumer to supplier and vice versa, think of a power feed-in using a photovoltaic system), such things can become even more complex.

The following figure shows an example load curve of a household on one day (right). The load duration curve (left) of a household in the following figure is often used for the analysis of load consumption in a certain time range (such as 4 hours in this case). The load curve gives the data as it is consumed in a household; the load duration curve arranges it in a way that gives an idea of the evolution from maximum to minimum.

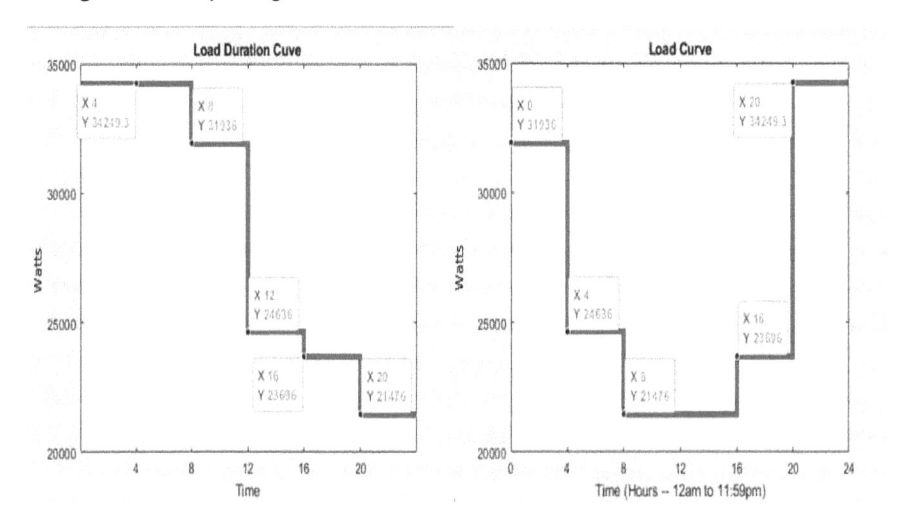

Figure 32: Load curve (right) and load-duration curve (left)

5.2 The power triangle

We have already learned a lot about capacitors and inductors in the previous section. Since such passive elements store energy, they create a phase shift between voltage and current - this also affects the power, which we have not discussed yet. We would like to do that in this chapter.

Consider an inductive circuit that draws a "lagging" current I and is connected to the source voltage U. The angle of this "lagging" is ϕ. If we assume that there is no phase difference between voltage and current ($\varphi = 0$ - purely resistive circuit), the total power P here is simply $U \cdot \mathrm{I}$. However, if we now add a reactive element (capacitor or inductor) to the circuit, a phase difference is created in the circuit ($\varphi > 0$). The total power in the circuit in this case is called apparent power:

$$S = U_{total} \cdot I_{total} \text{ (\textbf{Unit:} kVA)} \qquad 5\text{-}1$$

The apparent power is the sum (vector sum) of its components, one of which is the horizontal component (P), called **active power** (unit: kW) and the other being the vertical component (Q), called **reactive power (**unit: kVAr). Incidentally, the "VAr" here stands for VA reactive ("Volt-Ampère-réactif").

These three quantities (S and its components P and Q) form a triangle, which is called a **power triangle** (see Figure 33). Since the active and reactive power are components of the vector **S** is valid:

$$P = S \cdot cos\varphi = U \cdot I \cdot cos\varphi \qquad 5\text{-}2$$
$$Q = S \cdot sin\varphi = U \cdot I \cdot sin\varphi$$

The apparent power S is a vector consisting of the components P and Q, so that as for any vector for its magnitude:

Vector form:

$$\vec{S} = \vec{P} + \vec{Q} \qquad 5\text{-}3$$

$$S = \sqrt{P^2 + Q^2} = \sqrt{(S \cdot cos\varphi)^2 + (S \cdot sin\varphi)^2}$$

The phase difference between voltage and current can be expressed as:

$$\varphi = \tan^{-1}\frac{P}{Q} \qquad 5\text{-}4$$

Just as we learned with impedance that resistance (R) and reactance (X) are just the special cases of impedance (Z), we can also simplify to think of active power (P) and reactive power (Q) as special cases (actually components) of apparent power (S). Generally speaking, the active power is simply the power consumed by the ohmic components (e.g. resistor), and the reactive power is the power stored in the reactive components in the form of energy.

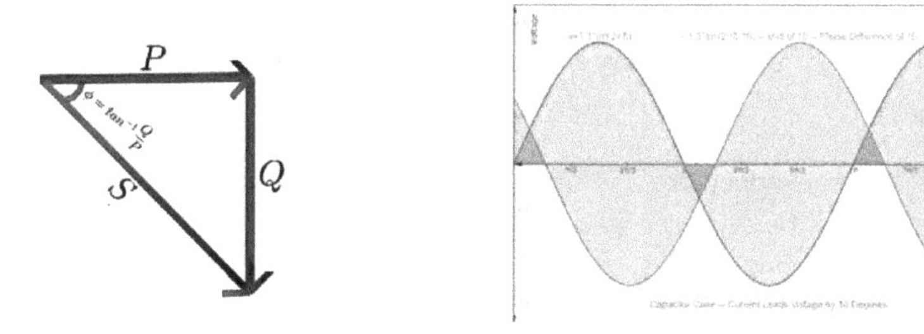

Figure 33: left: Power triangle - right: Phase difference (10 degrees) in the case of a capacitor

Power Factor:

Reactive power can be seen as a waste or derogatory by-product, so it should be minimal. This is measured by the power efficiency, which is called the power factor λ (also called power factor "PF" or "p.f."). Power factor is basically a measure of the quality of power in the circuit. No reactive power at all would mean perfect power quality. In mathematical terms, it means:

$$\lambda = \frac{|\vec{P}|}{|\vec{S}|} = \frac{S \cdot cos\varphi}{S} = cos\varphi \qquad \text{-5 5}$$

$$\lambda = \frac{|\vec{P}|}{|\vec{S}|} = \frac{I^2 \cdot R}{I^2 \cdot Z} = \frac{R}{Z}$$

Thus, the power factor is simply the cosine of the phase angle, and when the phase angle reaches zero, the maximum power quality ($\lambda = cos\varphi = 1$) is achieved, since in this case the reactive power ($S \cdot sin\varphi$) is equal to zero.

What does that mean? - Is not power equal to power?

For starters, this can leave you quite confused as to why power is not the same as power. Why is there a distinction made here between two powers, and why have they been given certain names? If active power is the power that the load requires for normal operation (produces heat, fairly conceivable), what role does this so-called reactive power play? These questions can be confusing at first, but it all makes sense if we look a little closer at the operating principle of reactive power elements.

Think back for a moment and remember that capacitors oppose a voltage change and inductors oppose a current change. The phase difference (φ) occurs primarily because of the opposing action of the reactive elements to such voltage and current changes. We already know that capacitors absorb voltage (charge) in the first quarter cycle and release it (discharge) in the second quarter. The same thing happens with inductors, but instead of voltage, they draw current. Since they take in (consume) and give out current, we can say that energy (or power) is stored in reactive elements in the form of fields. In the case of capacitors it is an electric field and in the case of inductors it is a magnetic field.

If we increase the inductance and capacitance of the electronic components, they can put up more resistance to change and the phase angle increases. As a special case for better understanding, let us consider the worst case scenario where the phase difference reaches 90 degrees (maximum). The only power the circuit has at this point is the reactive power ($S = 0,\ da\ cos(90) = 0$). Physically, it means that all the power here simply flows between the source and the load.

Now we already know the advantages of capacitors and inductors. In power supply systems, inductors are mainly used in electric motors, whose entire operation is based on rotation generated by the magnetic field of the inductor. Here, too, part of the energy is "wasted" in the form of reactive power. This must be mitigated as far as possible. To mitigate the effect of reactive power in the case of inductors, we use capacitors connected in series, for example. Why? You might figure it out if you think about phase shifting for a moment. We learned that the phase shift of an inductor is opposite to the phase shift of a capacitor. So if we get a reactive power of 1 kVAr from a source, an added capacitor of 700 VAr is already delivering 700 VAr to the inductor. So now the source only has to supply 300 VAr of reactive power to the inductor, thus "saving" 700 VAr. Adding capacitors to the inductive load in this way to reduce the reactive power drop is called **power factor correction. In** short, the goal here is simply to minimize the energy transactions from the reactive element and source as much as possible. These energy transactions occur because reactive elements store energy in the form of fields (magnetic and electric).

5.3 Single-phase and three-phase alternating current

Of course, AC systems are also not perfectly perfect, but still better for power transmission compared to DC. One of the advantages of alternating current over direct current is the possibility of three-phase system (also known as: heavy current, three-phase, power current).

A single-phase system (normal household current, three-core cable) basically only requires two wires: the **phase (L)** and the **neutral conductor (N)**. In addition, there is usually a protective **conductor (PE)** for protective earthing. The phase / outer conductor (L) (usually brown, black or red sheath) is a current-carrying wire. The neutral conductor (N), on the other hand, is usually sheathed in blue.

The current flows in the phase wire towards the load and after it has given up all its potential (voltage) to the load (or the load has taken it away from the current), it returns to the source via the neutral to get more voltage.

Three-phase (3ϕ) systems have the special feature that they have three phases or live wires (L1, L2, L3). However, they do not require three neutral conductors or protective conductors, but make do with one neutral conductor and one protective conductor each. Power cables therefore have a total of five cores (L1, L2, L3, N and PE).

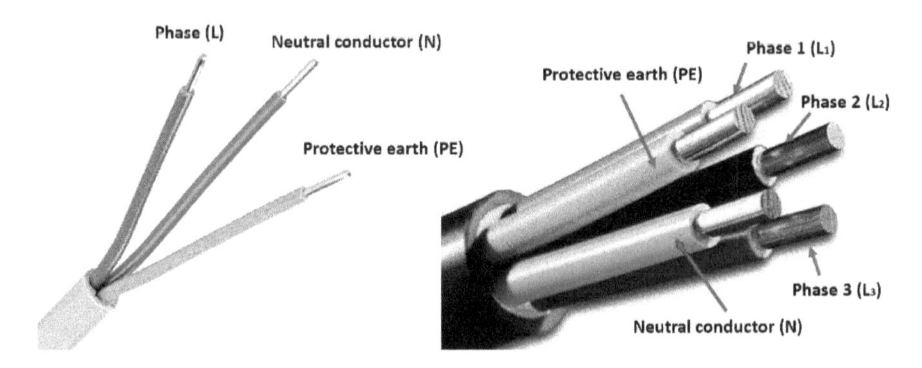

Figure 34: 3-wire (left) for single-phase current and 5-wire (right) for heavy current

The three currents are offset with a phase shift of 120 degrees. Three coils of a generator produce these three phases of the AC supply. Thus, in three-phase systems, the AC voltage can reach its peak value three times in one cycle. This characteristic of 3ϕ-systems creates a rotating magnetic field, which is necessary for the operation of AC machines.

With the phase angle of 120 degrees, we can describe the voltages of the three phases (A, B and C) as follows:

$$\begin{aligned} \boldsymbol{U_{AN}} &= U_P < 0° \\ \boldsymbol{U_{BN}} &= U_P < -120° \\ \boldsymbol{U_{CN}} &= U_P < -240° \end{aligned} \quad \text{5-6}$$

N stands here for the neutral conductor and U_P is the magnitude of the **phase voltage.** The phase voltage is formally defined as the voltage between the conductor and the neutral conductor. The current of a phase is called **phase current.**

Line voltage is the voltage between two phases.

$$\begin{aligned} U_{AB} &= U_{AN} - U_{BN} = U_P < 0° + U_P < -120° = \sqrt{3}U_P < 30° \\ U_{BC} &= \sqrt{3}U_P < -90° \\ U_{CA} &= \sqrt{3}U_P < -210° \end{aligned} \quad \text{5-7}$$

Equation 5-7 shows that the magnitude of the line voltage relates to the phase voltage as follows:

$$U_L = \sqrt{3}U_P \quad \text{5-8}$$

Using equation 5-7, it also follows that in the three phases, the sum of all voltages is zero, so: $U_{AB} + U_{BC} + U_{CA} = 0$.

Finally, by using KCL, we obtain that the sum of all currents (which is equal to the neutral current) is also equal to zero.

Balanced load:

A load connected to a three-phase AC system can be balanced or unbalanced. As the name implies, balanced means that all phases are equally loaded. Unbalanced, on the other hand, means that the phases are unequally loaded.

A symmetrical load can be configured in two ways: One is in star configuration (Y) and the other is in delta configuration (Δ) . In the star load, three phase wires are connected to a common star point and neutral. This configuration is used in current distribution.

The main feed from the power grid is connected to a Δ-Y transformer (delta-star transformer) which converts the Δ-feed into Y. Because of the neutral in Y, a single-phase feed with a neutral (common) can be supplied from these 3ϕ to loads (such as residential) with relatively low load requirements. Therefore, we have three wires in our everyday appliances and in the house (normal outlet), one of which is the neutral and the other is the phase. The third wire, as mentioned earlier, is the protective earth (PE) conductor for protective grounding. This protective earth conductor is sometimes not present (e.g. in simple plugs for lamps).

The line voltages in a star load configuration are different from the phase voltages (of a component), whereas the line current is the same as the phase current (since the load is connected in series). Just as we derived the line voltages earlier, the same strategy can be used here:

STAR CONFIGURATION

$$U_L = \sqrt{3}U_P \qquad \text{5-9}$$
$$I_L = I_P$$

A delta configuration, on the other hand, has only three phases connected in a ring and has no neutral. The delta configuration has its applications in voltage transmission. Apart from that, the delta configuration is used in electrical machines. The line voltage in the delta is equal to the phase voltage (since the load is connected in parallel):

DELTA CONFIGURATION

$$U_L = U_P \qquad \text{5-10}$$
$$I_L = \sqrt{3}I_P$$

Ordinary household consumers (e.g. hairdryer, washing machine, lamp) only require a single phase. So, from the three-phase supply of the power connection, a single phase with a neutral is distributed to areas of the house. All household loads have a common neutral that is used to balance the unbalances. To measure these unbalances, C. L. Fortescue developed the **balanced component** method, which states that each phase has three components, called zero, positive, and negative. This method is useful in solving problems with unbalanced loads. However, we will not discuss it in detail here.

Power plants generally try to maintain load balance, which means they forecast load demand and distribute the load evenly across each phase.

5.4 How does electricity get into the house? The power supply systems

In this section, we conclude the chapter with an overview of how the electricity is now distributed from the generation plants to the end consumers.

The main component of an AC power system is the transformer, which we will discuss in more detail in the next chapter. Transformers, simply put, only increase or decrease voltage. They have two sides, one is called primary and the other is called secondary. The power and impedance on both sides remain constant, so changing the voltage changes the current in the following way (Ohm's law): if you increase the voltage, the current decreases, which ultimately decreases the power dissipation. Now we could also transform direct current to high voltages, but due to various other advantages, alternating current systems are still preferable. A transmission and distribution network uses the following steps:

1) Electricity is (generally) **"generated"** at 11kV.

2) To reduce power losses, this 11 kV is then converted to 132 kV for transmission. From this point, the delta system (3ϕ, 3-phase) is used for **transmission.** 132 kV is an optimal choice because increasing it further would add more cost (e.g., wire insulation, switchgear, and other transformer equipment) and therefore would no longer provide an economic increase. If the voltage were to be increased further, the cost would increase more than the energy.

3) The **receiving stations** then step this voltage down to 33 kV and transmit the supply to the city's grid stations. This transmission is preferably carried out underground.

4) **The grid stations** further regulate this voltage down to 11 kV and distribute it via the transformers of the grid stations.

5) The transformers (Δ-Y transformers) then further reduce the voltage to 400 V and deliver it to the **consumer** with a neutral conductor. Some countries (such as North America and Canada) use 210 V systems; but 400 V is still the most common. Depending on the country, 230 V or USA: 110 V is then present at the domestic socket.

Poles and towers (for 132 kV supply) are used for the overhead lines (see Figure 35).

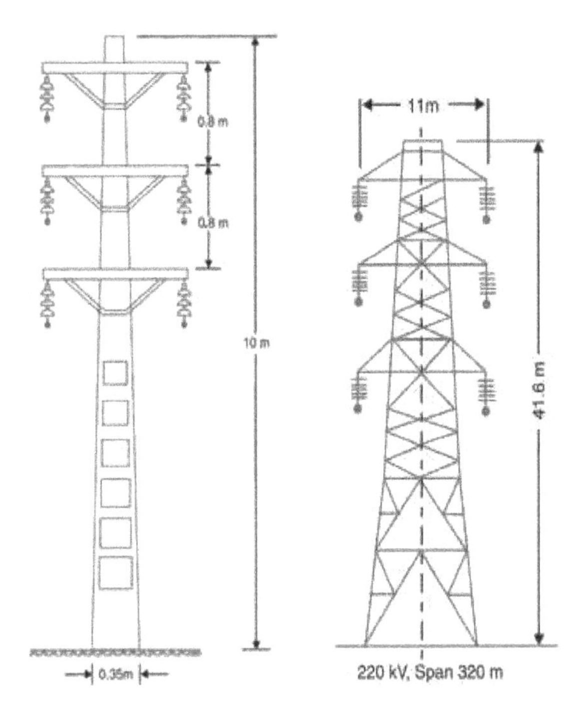

Figure 35: Transmission tower (left) and steel tower (right)

5.5 Protection of the power grid

Various faults can occur in the power supply system that can cause the entire distribution system to shut down and sometimes cause severe damage to expensive equipment. Therefore, it is important to minimize the occurrence of these faults and when they do occur, it is important to have a system that protects the expensive equipment and restores power in the shortest possible time.

Reasons for these short-circuit faults can be, for example, insulation damage, a lightning strike, a specific accident or even natural disasters. To protect the system, relays with circuit breakers (CBs) are used to prevent current surges. The relay in protection systems works in different modes with current and potential transformers (CT and PT). Current transformers reduce the amperage to a lower level (usually 5 A standard), and potential transformers reduce the voltages. A current transformer with 5 A secondary is normally connected to a relay device that trips the circuit breakers in the event of a fault.

The detection devices in fault detection are therefore relays. The quality of a reliable relay is that it must guarantee reliability (the relay works on all faults) and safety (the relay does not work on false faults). Relays are designed based on these two parameters. The setting of the so-called pick value of the relay (voltage limit for relay operation) is for reliability and the setting of the time delay (delay after the relay CB signals drop out) is for safety.

In addition to traditional mechanical electromagnetic relays, there are now also solid-state relays with feedforward control and autonomous operation. These relays can be used in concepts such as a so-called smart grid. The problem with faults is that once they occur, they usually require human intervention. In order to reduce human (decision) errors, this concept of autonomous relays, comparable to the concept of autonomous driving, offers enormous potential for improvement. However, implementation is (still) limited due to resource and complexity constraints.

6 Electrical machines

Electricity has become a main source of energy for us, as the energy we need for our daily tasks and entertainment (working, making coffee, doing laundry, watching TV,...) can be obtained by converting electrical energy into other forms. When we do mechanical work (e.g. drilling a hole with a drill), this requires the conversion of electrical energy into some kind of rotational energy. Electric motors are the basis for this. Converting mechanical energy back into electrical energy is again possible with the help of generators. Basically and simplified, by the way, every electric motor can be used as a generator and every generator can be used as an electric motor at the same time, because they have the same design. It only depends on whether current is connected or whether the shaft is turned by mechanical work and then current is tapped at the connections.

In this chapter we will cover the physics, basic principles and operation of electrical machines. The chapter is intended for a basic understanding of electrical machines. By the way, we will only cover electric motors in the course of the chapter, since generators are identical in construction, as already mentioned.

6.1 Magnetic field and electrical machines- basics

6.1.1 Faraday's law of induction

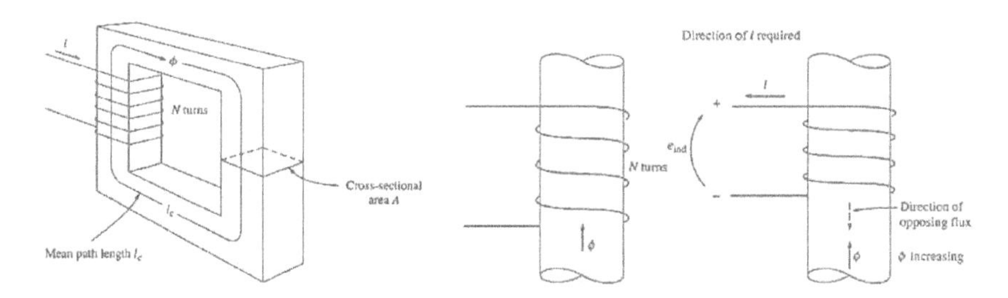

Figure 36: Ampere's law (left) ; Faraday's law (right).

As we have already learned, a current-carrying conductor wound around a magnetic core produces a magnetic field *(***Ampere's law)**. Similarly, we already know that the change of the magnetic field induces a voltage in an inductor (**Faraday's law)**.

A current-carrying wire that generates a magnetic field has a certain **magnetic flux** Φ. This magnetic flux Φ depends on the number of turns of the coil (n) wound around a magnetic core of length (l_c) and area (A). The more magnetic the material used for the core, the more current this flux produces. This ability of the magnetization of a material is called its **permeability** (μ).

Ampere's Law

$$\Phi = \mu \cdot \mathrm{I} \cdot \frac{\mathrm{n} \cdot \mathrm{A}}{l_c} = k_1 \cdot \mathrm{I} \Rightarrow \Phi \propto I \quad (6\text{-}1)$$

For example, the permeability of steel is about 1600 times greater than that of air, so magnetic flux in steel induces 1600 times more current than magnetic flux in air.

For electrical machines, we are mainly concerned with the **flux linkage** (also induction flux, linkage flux, coil flux) ψ, which is defined as the total magnetic flux of a coil (inductor). This is obtained by integrating the magnetic flux density over the area of the coil including connections. For a homogeneous field (field lines of equal strength and direction) in the core, however, this can be expressed in simplified terms for a coil with n turns as follows:

$$\psi = n \cdot \Phi = \mu \frac{n^2 A}{l_c} \cdot \mathrm{I} = L \cdot I \quad (6\text{-}2)$$

If we make a comparison with the capacitor, where we had the capacitance (C) depending on the dielectric medium (ε), the distance between the plates (d) and their area (A), here we have the inductance L of a coil depending on "n", "A", "l_c" and "μ".

It therefore applies:

Faraday's law of induction

$$\frac{d\psi}{dt} = U_{ind} = -n \cdot \frac{d\Phi}{dt} \quad (6\text{-}3)$$

6.1.2 The magnetic-ohmic law

When electric current flows through a coil, it induces a magnetic field in the core. The lower the magnetic resistance (**reluctance**; R_m) of the core, the more electromotive force F, which in turn increases the magnetic flux (Φ) is generated by the current. Just like the electrical resistances in a series circuit, the reluctance of the magnetic circuit adds up $R_{m1} + R_{m2} + \dots + R_{mn}$ and just like in the parallel electrical circuit, the following expression is obtained: $(\frac{1}{R_{m1}} + \frac{1}{R_{m2}} + \dots + \frac{1}{R_{mn}})^{-1}$.

6.1.3 Force in a current-carrying conductor in a magnetic field

When we place a current-carrying conductor in a magnetic field of density **B** (**magnetic flux density**), this induces a force, called the **Lorentz force**, on that conductor. The reason for this force is simply the interaction between the conductor and the surrounding magnetic field.

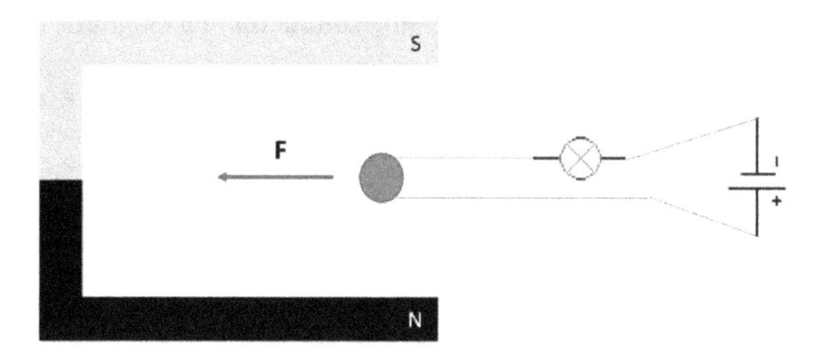

Figure 37: A current-carrying conductor (blue) experiences a (Lorentz) force in a magnetic field

The vector of this force is:

$$\mathbf{F} = I \cdot (\mathbf{l} \times \mathbf{B}) \rightarrow F = I \cdot l \cdot B \cdot sin\theta \quad \text{6-5}$$

$$F \cdot r = \tau = r \cdot \mathrm{I} \cdot l \cdot B \cdot sin\,\theta \quad \text{6-6}$$

The "r" here is the radius of the loop and is needed to get the torque equation and "B" here is the magnetic flux density.

In addition:

$$\Phi = B \cdot A \quad \text{6 -7}$$

To determine the direction of the force or force vector, there is what is known as the right-hand rule, which you may have heard of. If you find it difficult to understand the formulas above, just imagine that the force here is perpendicular to a plane I x B ($\mathbf{F} \perp \mathbf{I} \times \mathbf{B}$). So, since the force is a perpendicular component of the plane I x B, according to the law of triangles, you can use $sin\ \theta$ can be used. Since the current is not a vector quantity, we will take its length (I) as a vector here. As a reminder, for the **cross product of** two vectors, the cross product of two vectors that are perpendicular to each other yields a new vector that is perpendicular to the two initial vectors.

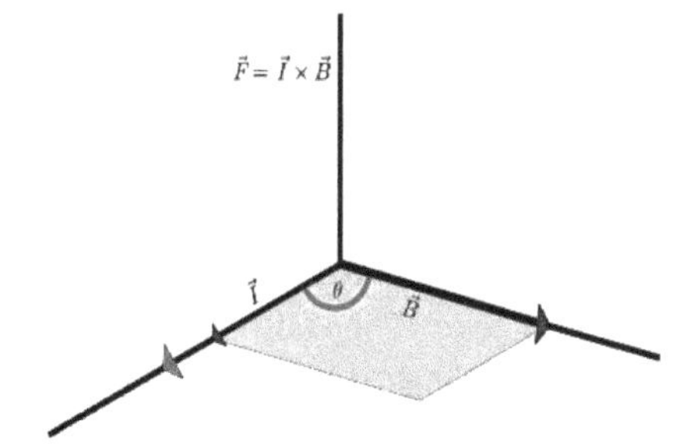

Figure 38: Vector multiplication of two vectors, shown here in the 3D plane

6.1.4 Voltage in the current-carrying conductor in a magnetic field

We already know that a change in the magnetic field induces a voltage in a conductor and thus a current starts to flow in it. Let us now consider a conductor which is in a magnetic field and in which a current I flows. The current produces a magnetic field, which then interacts with the surrounding magnetic field, which in turn produces a change in the magnetic field. This change in the magnetic field then induces a voltage in this conductor with the value:

$$\boldsymbol{U}_{\mathbf{ind}} = (\mathbf{v} \times \mathbf{B}) \cdot \mathbf{l} \Rightarrow U_{ind} = v \cdot B \cdot sin\theta \cdot l \cdot cos\theta \qquad 6\text{-}8$$

6.1.5 Torque in a current-carrying loop

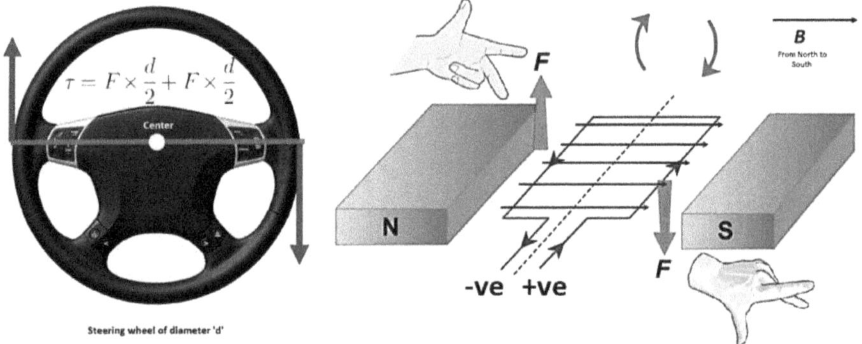

Figure 39: Pair of forces in a current-carrying loop in a magnetic field

If we place a current-carrying conductor (rectangular shape; coil) in a magnetic field as shown in the figure, this magnetic field generates a torque (as in mechanics: e.g. when steering with a steering wheel, see figure), which causes the conductor to rotate in this magnetic field. Using the example of the steering wheel, simply imagine the steering wheel as a circle with a certain diameter d.

When we steer to the right (i.e. turn the steering wheel clockwise), we apply an upward force on the left side of the steering wheel and a downward force on the right side with our hands. So the force vectors are in opposite directions. The same principle is used in the case of an electric motor. Here the density vector (**magnetic flux density B**) of the surrounding magnetic field is fixed. So if the direction of the conducting wire (vector **L**) changes, the force (**F**) will also change. When both conductors align with the magnetic field, the forces cancel out. (Example steering wheel: if you take your hands off the steering wheel, it will align itself in neutral). Mathematically:

$$\tau = \frac{d}{2} F sin\theta + \frac{d}{2} F sin\theta = dF sin\theta = 2rF \sin\theta \qquad 6\text{-}9$$

Here $\frac{d}{2}$ (= r) is simply the lever arm (radius) from the fulcrum (centre) of the conductor loop (circle; in the example this would be the centre of the steering wheel). If you now insert "F" from equation 6-5 and convert it, you get a simple, beautiful form that can describe the entire operation of an electric motor. This following equation describes the torque (τ) of a current-carrying coil in an external magnetic field:

$$\tau = N \cdot I \cdot A \cdot B \cdot sin\theta \qquad 6\text{-}10$$

θ: Angle between field lines and a perpendicular to the plane of the coil.
I: current
A: area of the coil
B: magnetic flux density
N: coil windings

Voltage in a current-carrying loop:

To obtain the voltage induced in this loop, we can follow equation 6-8. Following the same processes, we can make an equation analogous to the above:

$$U_{ind} = 2vBL \sin\theta \qquad 6\text{-}11$$

Which, after rearranging, yields:

$$U_{ind} = \Phi_{max}\omega \sin\theta \qquad 6\text{-}12$$

6.2 Transformers

Transformers are basically electrical machines that follow the principle of **electromagnetic induction**, more precisely mutual induction (also: **mutual induction**, self-induction, inductive coupling). A transformer usually consists of two (or even more) coils, which are placed relatively close to each other, e.g. on a common magnetic core (iron).

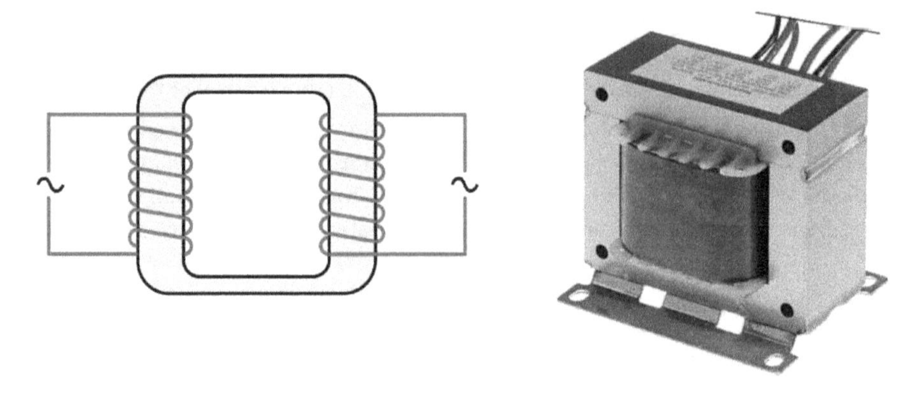

Figure 40: Schematic principle of a transformer (left) and real example of a transformer (right)

In simple terms, this mutual induction in a transformer (transformer) then works as follows: An alternating voltage (~) and the associated alternating current generates a magnetic field in the vicinity of the first coil (primary side). This in turn generates (induces) a voltage in the second coil (secondary side). The constantly changing alternating current in the first coil thus induces a voltage in the second coil. This is also an alternating voltage and has the same frequency as the first voltage.

If we connect a voltage source U_P (primary side) to coil one, this therefore induces a voltage U_S (secondary side) in the second coil. The current that these voltages produce depends on the inductance (L) of the coils and hence the number of turns (N) of both the coils. We can therefore say that the ratio of the number of turns of two coils is equal to the ratio of the two voltages.

VOLTAGE

$$\frac{U_P}{U_S} = \frac{N_P}{N_S} = a \qquad \text{6 -13}$$

Here, "a" stands for the **transmission ratio of the transformer**.

Transformer power can be defined by Equation 5-2 as:

$$P_{in} = U_p \times I_p \cos \varphi \text{ and } Q_{in} = U_p \times I_p \sin \varphi$$
$$P_{out} = U_s \times I_s \cos \varphi \text{ and } Q_{out} = U_s \times I_s \sin \varphi \qquad \text{6-14}$$

Since the power in both coils of the transformer is the same:

$$P_{in} = P_{out} \Rightarrow U_P \cdot I_P = U_S \cdot I_S \qquad \text{6 -15}$$

6.3 Direct current machines (direct current motor)

When we connect a DC source to the rectangular coil in Figure 39, current (I) begins to flow. The flow of this direct current in this rectangular coil, which is in a magnetic field, induces a pair of forces on both sides. Since this force now causes the coil to rotate 180 degrees, the direction of the force pair shifts when the polarity of the terminal voltage is the same. In Figure 39, the +ve terminal produces a downward force and the -ve terminal produces an upward force. So if we swap these terminals and use the right-hand rule again, this reversal of the current shifts the direction of the force. Because of this shift, the motor reaches a state of equilibrium after completing this 180 degree cycle (as in. $\sum \tau = 0$; remember what happens when you let go of the steering wheel), and will not turn any further.

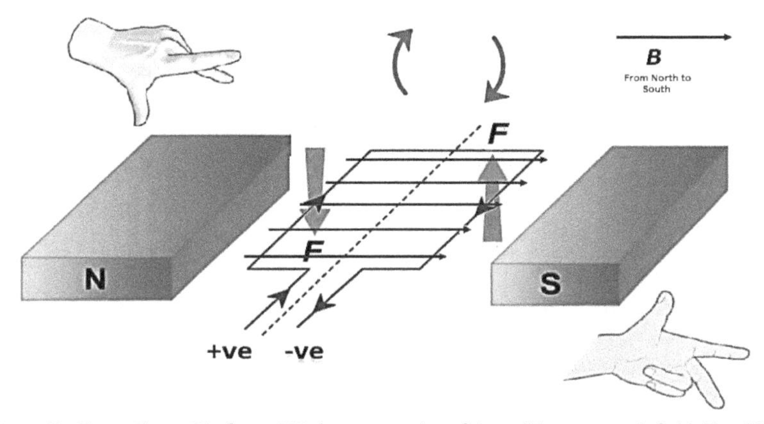

Figure 41: Shows Figure 39 after a 180 degree rotation of the coil in a magnetic field. The shift in terminal voltage shifts the direction of the force.

To solve this problem, a sophisticated design called a **commutator** is used. Namely, for the continuous motion of a DC motor, we need a component that automatically swaps the positive and negative voltages of the connected DC power supply. We have already recognized this in the previous paragraph. The **commutator** is simply a component that creates a pole change in the rotating part (**rotor**). This is done by means of brushes (**carbon brushes**) which are connected as fixed contacts and loop on a ring which has two interruptions. It is therefore basically nothing more than a component that temporarily interrupts the flow of current.

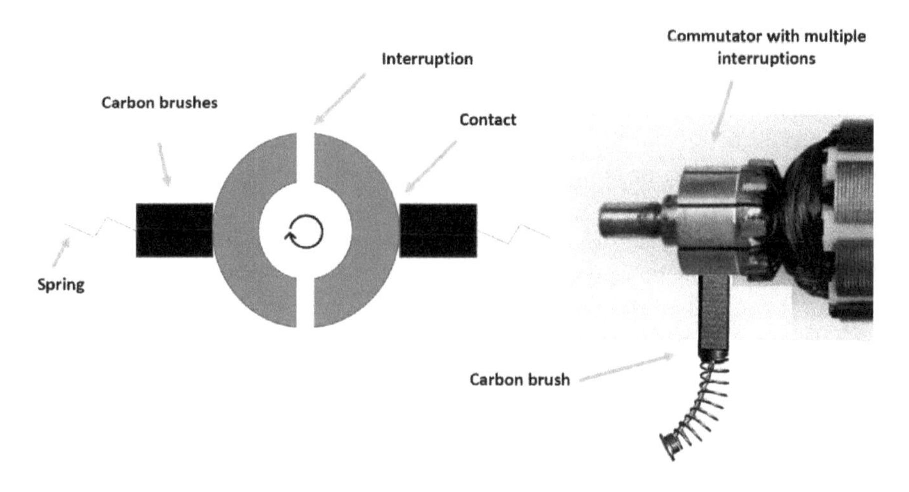

Figure 42: Commutator (left: schematic , right: real) of a DC motor

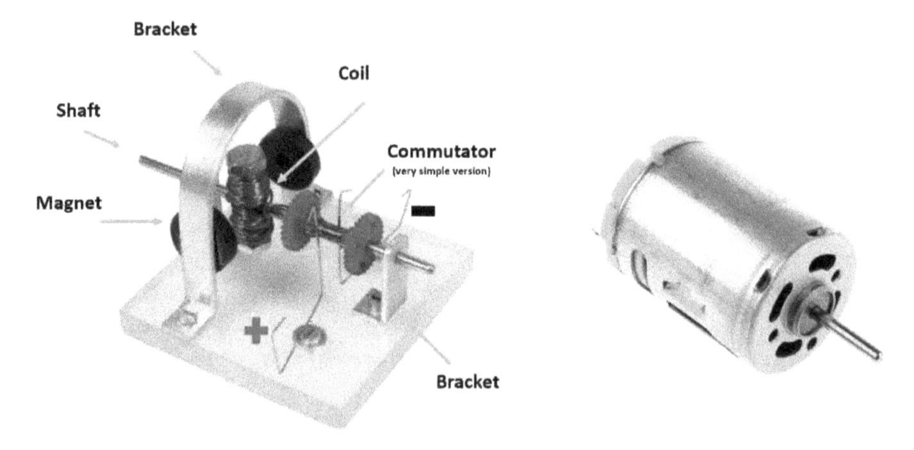

Figure 43: DC electric motor (left: schematic, right: real)

These illustrations refer to two-pole machines only, which physically simply means that there are only two slots in the motor where loops are wound. You can think of it like this. For example, take a copper wire and wind it around two rods. Start the winding on the first rod and finish the winding on the second rod. However, we could also wind the wire around three rods, start with the first, wind some copper on the second, and then end the winding on the third. The structure of an electric motor contains slots (instead of rods) where these windings are wound. These windings are commonly referred to as **armature windings**. The armature winding is placed in what is called the **rotor**, which is the rotating part of the electric motor. A **shaft, which** can then be attached to this rotor, is used for the output, i.e. the use of this mechanical energy for various purposes. In the case of a drilling machine, for example, from a mechanical point of view, the drill chuck would be connected here, in which in turn a drill can be clamped, which then rotates. As well as a two-pole structure, as mentioned earlier, there are now not only two-pole motors but also three-pole. In a three-pole machine, in contrast to the diagram of the commutator (left side in Figure 42), we would find here three slots or interruptions in the commutator. A three-pole DC machine is commonly used and many simple motors are often already three-pole.

Sometimes, instead of the permanent magnet (**stator**) located in the core, a winding / coil is also installed. This field winding / coil behaves like an electromagnet and thus provides a surrounding magnetic field to this current-carrying conductor. The motor with a permanent magnet is generally referred to as a permanent magnet DC motor (PMDC) and the one with a field winding, i.e. a coil in the core, as an electromagnetic motor.

6.3.1 Analysis of circuits with DC motors

Solving circuits with direct current (DC) motors is relatively simple, since they are known to contain only a rotor and a stator. Circuits with permanent magnet DC motors (magnet in the core) are not so complex, since the magnetic field of the permanent magnet is constant and can thus be treated as a simple constant. To vary the torque and speed of such a DC motor, we can simply affect / change the current in the field winding. The equivalent circuit of the rotor contains a simple voltage (U_R), and its resistance (R_R) - and the field part contains the resistance (R_F) and the inductance (L_F).

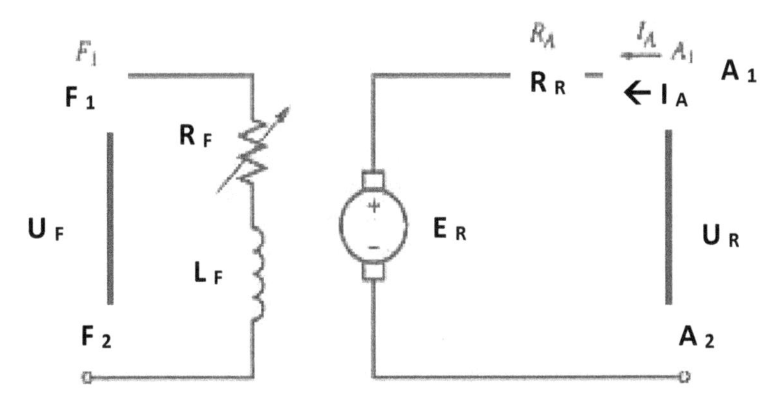

Fig. 6 9: DC motor

The terminal voltages can be easily released by KVL. The voltage U_R here, as we know from equation 6-8, is simply the voltage induced across the coil. This equation can be converted to angular velocity form as:

from equation 6-8: $$U_A = vBl = r\omega Bl = rlB \cdot \omega = AB \cdot \omega \quad \text{6-17}$$

from equation 6-7: $$U_A = \Phi\omega$$

The torque can be derived in the same way as from equation 6-6 as:

from equation 6-6:

$$\tau = rI_A lB = rlB \cdot I_A = AB \cdot I_A \qquad 6\text{-}18$$
$$\tau = \Phi I_A$$

In the same way, we can define the power of a DC motor (similar to the power of an internal combustion engine in "horsepower") as:

$$P = \frac{W}{t} = \frac{Fd}{t} \qquad 6\text{-}19$$

$$v = \frac{d}{t} = r\omega \Rightarrow$$

$$P = \frac{\frac{\tau}{r} \cdot (vt)}{t} = \frac{\tau}{r} \cdot (v) = \frac{\tau}{r} \cdot (r\omega)$$
$$P = \tau\omega$$

6.4 Alternating current machines (alternating current motors)

6.4.1 Basics of AC motors

We already know that a magnetic field induces a force or torque in a current carrying loop. Let's take another look at Figure 39, here the force on the left conductor of the loop is directed upward because the north pole of the permanent magnet is not aligned with the magnetic field of the current carrying conductor. Since two like poles face each other, they repel each other, which creates a torque in the conductor. This torque then moves the left conductor of the loop toward the south pole of the permanent magnet. In this position (Figure 41), with the north pole of the conductor facing the south pole, the loop reaches equilibrium and the motion stops ($\sum \boldsymbol{\tau} = \mathbf{0}$). From this we have concluded that after a 180 degree turn the movement of the motor stops (Figure 41), because when the current changes, the forces also change.

In summary, we can say that the magnetic field of the conductor (**rotor**) always tries to match the magnetic field of the external magnet (**stator**). We could also say that the north or south pole of the rotor always follows the south or north pole of the stator.

To solve this problem, DC machines use the commutator, which reverses the direction of the current after each 180 degree rotation of the loop. We already know this. But now there is another case where the need for a commutator does not exist. Namely, if by some means we can make it possible to rotate the magnetic field of the stator. Then the rotor will constantly follow it and the turning movement will not stop.

6.4.2 A rotating magnetic field

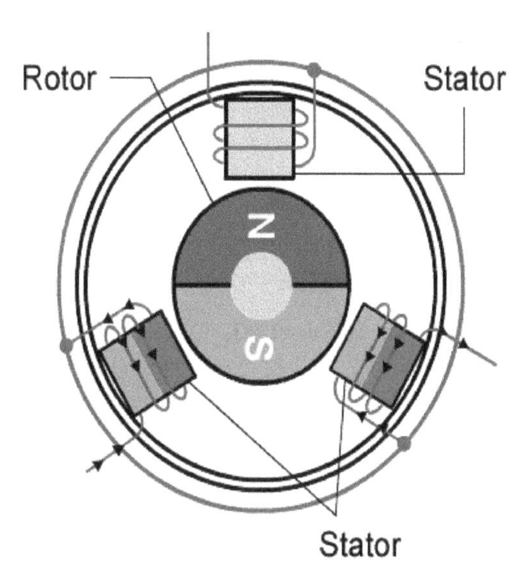

Figure 45: Operating an electric motor with alternating current

In the previous sections we learned that when we pass direct current through the windings of the stator, a torque is induced in the loop which, without a commutator, exists only for one 180 degree revolution. But if we now run alternating current in instead of direct current, the magnitude of the electromagnetic field varies sinusoidally, which means that in one cycle the north and south poles alternate as the sine wave travels from the positive to the negative peak ("uphill and downhill"). After a 180 degree rotation, the electromagnetic field of the stator under alternating current swaps the poles and so the forces acting on the rotor also shift. In order to swap the poles, we do not need a special component such as a commutator here under alternating current, since the property of the alternating current virtually replaces this component and shifts the forces itself.

Due to the sinusoidal nature, the motor accelerates in the first half quarter cycle and then slows down again in the second quarter (sine wave). To get around this, we can use three-phase AC. In the three-phase system, as we already know, we have a shift of 120 degrees in each phase, and if we now fix these phases at 60 degrees apart (360 degrees / 6 = 60 degrees), the direction of this magnetic field changes, but the magnitude does not. With three-phase alternating current, the direction of the magnetic field changes (it rotates), but its magnitude remains the same. The addition of the magnetic fields of the three currents of the three-phase system in different positions would also prove this effect mathematically. However, we will dispense with this proof here.

6.4.3 AC motor types

There are two types of AC motors: synchronous **machine** (or synchronous AC motor) and **induction machine** (asynchronous). In the synchronous **machine (SM),** the stator and the rotor run synchronously (with respect to the rotating field). When the magnetic fields of the rotor (generated by the current carrying loop) and the stator interact, the rotor starts "chasing" the electromagnetic three-phase field of the stator and eventually catches up with it (synchronization).

The other variant is the **induction machine (IM)**. In synchronous machines we have two magnetic fields, one coming from the current carrying loop (rotor) and the other from the three phase AC voltage (stator). In the induction machine, however, the electric current in the rotor is generated (induced) by the magnetic field of the stator coil. Let's take a closer look. By now we already know the concept of mutual induction and Lenz's rule. Now, when we apply three-phase AC current to the stator windings of the induction motor, this induces a voltage in the second coil (rotor coil). So instead of generating a special magnetic field by applied current in the rotor, as is the case with synchronous machines, here the stator simply induces some of its energy into the rotor, thus generating the current in the rotor. This magnetic field of the rotor induced by the stator always acts in the opposite direction to the magnetic field of the stator (Lenz's rule) and thus ultimately produces the same effect as a synchronous motor. The only difference is that here the rotor never catches up with the rotating magnetic field of the stator and thus there is always a so-called **slip** between them. The slip here simply indicates the difference in speed of the rotor and stator. The slip of synchronous machines, on the other hand, is always zero, i.e. non-existent.

Why three-phase?

If we were to run an induction motor on single phase AC, it would produce a fault due to this slip and stop after a few revolutions. The movement of the motor stops when the magnetic field of the rotor aligns with that of the stator. The slip here sooner or later creates a problem in the magnetic field alignment. To solve this, we use three-phase AC. We can also use two-phase AC, with each phase 180 degrees apart and 90 degrees away (360 degrees / 4 = 90 degrees). In this case, if the coil is placed vertically, the coil will be attracted to the horizontal phase component when we turn the motor on. So the position of the rotor does not matter, one phase always attracts it (see Figure 46).

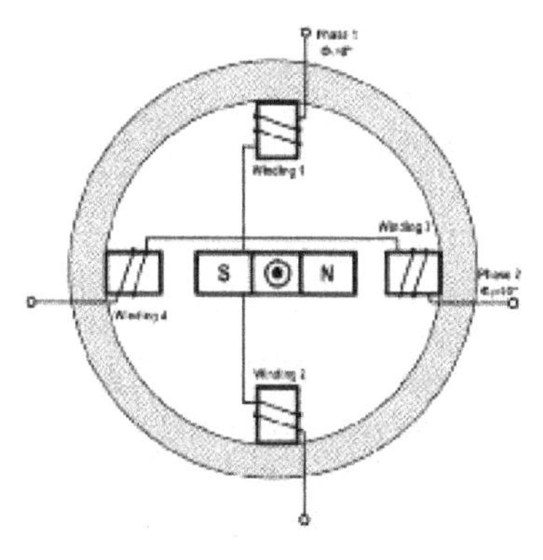

Figure 46: The position of the magnet (rotor) does not matter, a winding always attracts it

How does it still go single phase?

A simple ceiling fan, for example, can operate on single-phase alternating current because it contains a capacitor (usually 2.5 μF). If this capacitor is defective, you will notice that the fan does not start automatically. This is because capacitors here simply create a 90 degree phase shift, in a sense creating a two phase condition on this single phase fan. So each phase in this case is separated by 90 degrees (Figure 46) and the motor can start by itself. The mathematics of AC machines is beyond the scope of this book, but there is an important equation to know. This equation relates the frequency of alternating current to speed. As we increase the AC frequency, the rotors move faster (since there are now more 360 degree rotations in a cycle - at 50Hz, for example, there are 50):

$$rotor\ speed = \frac{120 \times \text{frequenzy stator (AC)}}{Number\ of\ poles\ in\ the\ machine} \cdot (1 - Slip) \qquad 6\text{- }20$$

7 Renewable Energies

Energy is a basic need for the growth and maintenance of a civilization, as work requires energy. Global electricity consumption for the year 2014 was approximately 726.6 MWh. Almost 60% of this energy was generated from fossil fuels, resulting in a total carbon footprint of approximately 35.25 billion tons. This amount of carbon dioxide contributes to the global climate crisis and can cause temperatures to rise, further melting glaciers and raising the level of the world's oceans. The depletion of the ozone layer is allowing higher levels of UV radiation into the earth's atmosphere, which poses an increased health risk (skin cancer, etc.) to us humans. It is likely that the world population will increase to about 10 billion people by 2050. The more people and the more complex our technology, the more electricity and energy we will need. A large supply of **green energy** is important for the environment as well as for us humans for the reasons mentioned above.

Due to economic reasons, it is difficult to simply eliminate energy production from fossil fuels, as energy is now an integral part of our progressing society and the existing generation structure would have to be completely changed. We still need to find better and better methods in terms of renewable energy sources to minimize further impact on the environment and our future.

In this chapter we will cover two of the better known renewable energy sources, namely photovoltaic systems and wind turbines for electricity generation. There are also other renewable energy sources, such as hydroelectric plants, hydrogen fuel cells, biomass and geothermal energy. But apart from hydropower, these methods have efficiency problems and other difficulties. Some of them, such as geothermal, work only in certain areas with hot springs. So in this chapter, we will only look at the two most popular renewables that can help us minimize carbon emissions on a large scale.

7.1 PV systems -photovoltaics

The irradiance (unit: $\mathrm{W/m^2}$) is the measure of solar irradiance received per unit area. Different regions on earth have different irradiance levels (see Figure 48). PV systems are particularly worthwhile in regions with medium to high irradiance in order to achieve maximum efficiency in electricity generation.

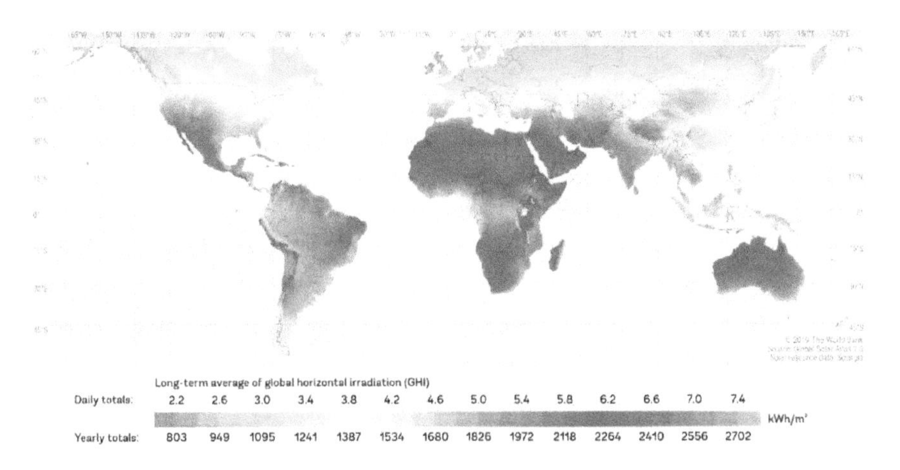

Figure 48: Irradiance on a world map

Figure 48 shows that the earth receives a maximum of 7.4 kWh/m2 of irradiation in one day. The maximum efficiency of solar modules in 2021 is about 19 %, which would result in almost 1.41 kWh of energy per square meter in one day.

How does photovoltaics work? Where does the electricity come from?

Photovoltaic modules are usually made of silicon. When sunlight hits such a photovoltaic cell (see Figure 49), the energy that the sunlight brings with it is converted into electricity in the cell. The background or basic principle for this is called the "**photoelectric effect**".

This effect describes the process of dissolving electrons from a semiconductor surface (metal surface also possible) under the incidence of light (**photons**). Electrons are thus released and transported further due to a special doping of the semiconductor element.

The generated electrical voltage (DC voltage) can then be tapped at the connections of the photovoltaic system. In order to use this voltage, an inverter is needed to convert the DC voltage into AC voltage so that the "generated" electricity can be fed into the power grid.

PV system design:

A **photovoltaic cell** consists of crystalline silicon, which converts light energy into an electron-hole pair and thereby generates 0.5 V. A voltage of 18 V is obtained by connecting this cell 36 times in series. This pair of 36 cells is called a **module.** To further increase the voltage, the combination of modules in series forms what is called a **PV string**. And this combination of strings in turn forms a **PV array**, which we can then install on our house roof, for example.

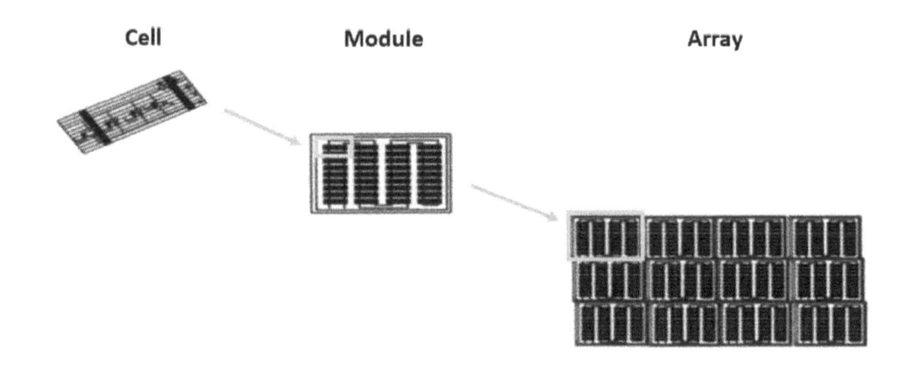

Figure 49: The structure of a photovoltaic array

Each individual **PV module** has a maximum power point (like a module with constant 18V). If the short circuit current at 0 V is equal to I_{sc} and the open circuit voltage at 0 A is equal to V_{oc} then the maximum power is in between. If we increase the current above 0 A, the voltage increases V_{oc} increases, and in the same way, when we increase the voltage above 0 V, the current decreases I_{sc}. The central **optimum point where** the product of the two values gives the maximum value is called the **point of maximum power "Maximum Power Point"** ($\boldsymbol{P_{MPP}}$).

$$P_{mpp} = U_{mpp} \cdot I_{mpp} = FF \cdot U_{oc} \cdot I_{sc} \quad \text{7-1}$$

Here FF (fill factor) is a constant for a PV cell that describes how high the efficiency of a PV module is.

Sample example 5

Design a PV system for a home that requires 10 kW at 220 V, 50 Hz single-phase AC. U_{mpp} and I_{mpp} of the PV panels are 54 V and 3 A.

Here we will take the input from the solar panels and feed it into the inverter. Inverters take the DC supply (U $_{DC}$) and convert it to AC (U $_{AC}$) by quickly switching the DC polarity. The equation for the output voltage of the inverter is:

$$U_{AC} = \frac{U_{DC}}{\sqrt{2}} \cdot 0,9 \Rightarrow \mathrm{U_{DC}} = \frac{220}{0,9} \cdot \sqrt{2} = 345,6\ V$$

To get this 345 volts from solar cells, we need:

$$\text{Number of modules} = \frac{345\ V}{54\ V} = 6,4 \approx 7\ \text{modules}$$

Each module can generate 54 volts. Connected in series, these 7 modules generate:

$$\text{Voltage} = \mathbf{7 \cdot 54\, V = 378\ V} \quad \Rightarrow \quad \text{Power} = \mathbf{378\, V \cdot I_{mpp}} = \mathbf{378\, V \times 3\, A} = \mathbf{1134\, W}$$

So to achieve the required power of 10 kW, we need:

$$\text{Number of PV strings} = \frac{\mathbf{10\,000\, W}}{\mathbf{1134\, W}} = \mathbf{8{,}8 \approx 9}$$

So here we can use two PV arrays with a total of 5 PV strings:

$$\textbf{Array-Power} = \mathbf{5 \cdot 2 \cdot 1134\, W \rightarrow 11{,}34\, kW}\ \textbf{Power}$$

7.2 Wind Turbines

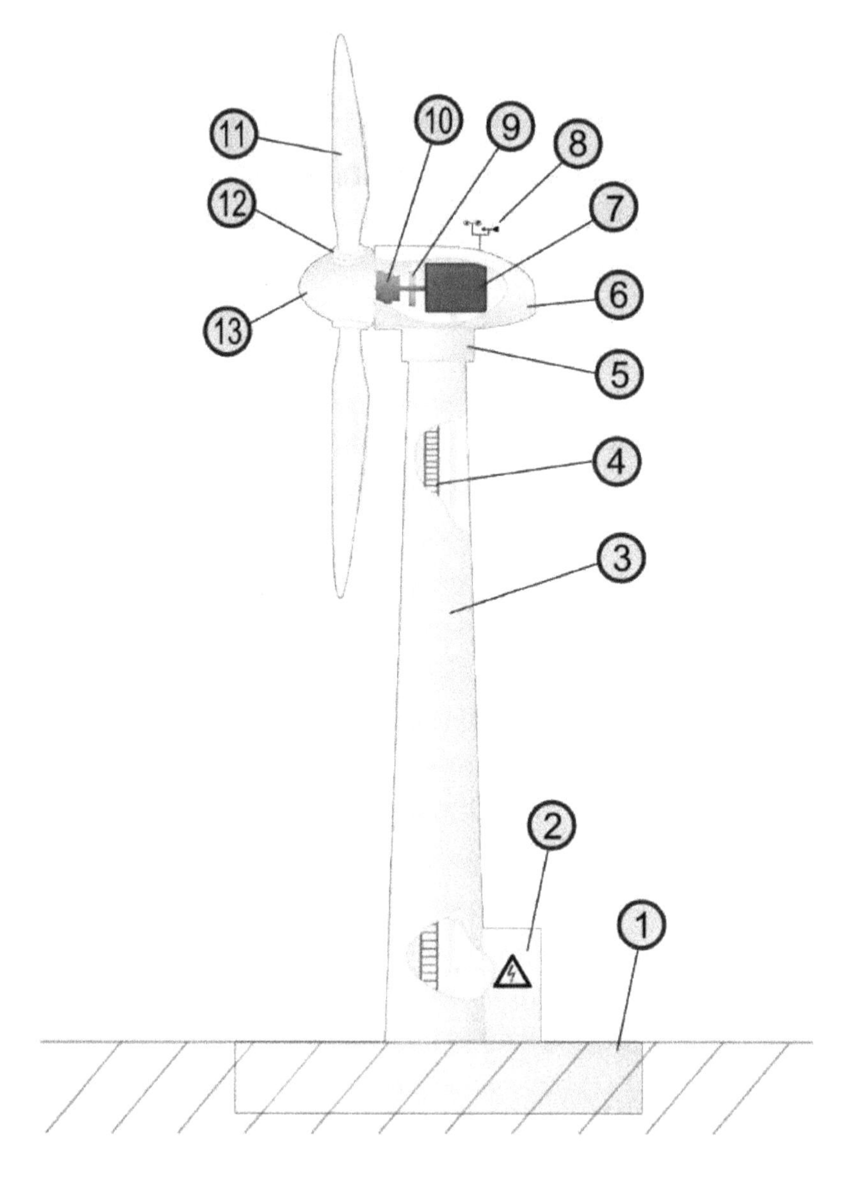

Figure 52: Structure of a wind turbine (Copyright: Arne Nordmann (norro) - CC-License)

1) Foundation 2) Connection to electric grid 3) Tower 4) Access ladder 5) Wind orientation control 6) Nacelle 7) Generator 8) Anemometer 9) Brake 10) Gearbox 11) Rotor blade 12) Blade pitch control 13) Rotor hub

Wind turbines use **synchronous generators**, where the energy generated by the wind turns the generator rotor, which induces a voltage in the stator. The **gearbox** increases the speed of rotation of the generator rotor. Since the radius of the generator rotor is smaller inside than the **turbine blades**, the design of the turbine requires proper calculation of the gearing to keep the total torque the same. Since wind blows in different directions, modern turbines use pitch and yaw motion control. Pitch motion in turbines applies to the blades and yaw motion applies to the entire turbine (including the generator and gearbox). Other parts of the turbine are important for the casing, weather protection, speed control, but are not essential for the functional operation of the wind turbine.

To calculate the voltage generated by the wind turbine, we need to consider the gear ratio. If we want a generator to run 8 times faster than the blades we have to design a gear ratio of 1 : 8 in which we have 4 blade gears and 48 generator gears. Normally, generators in wind turbines operate at a speed of 120 rpm. Now when the rotor turns, it generates electricity in the turbine which is then transmitted to the power grid for distribution. As noted with the electric motor, the generator is not really different in design.

Appendix A: Overview of circuit symbols

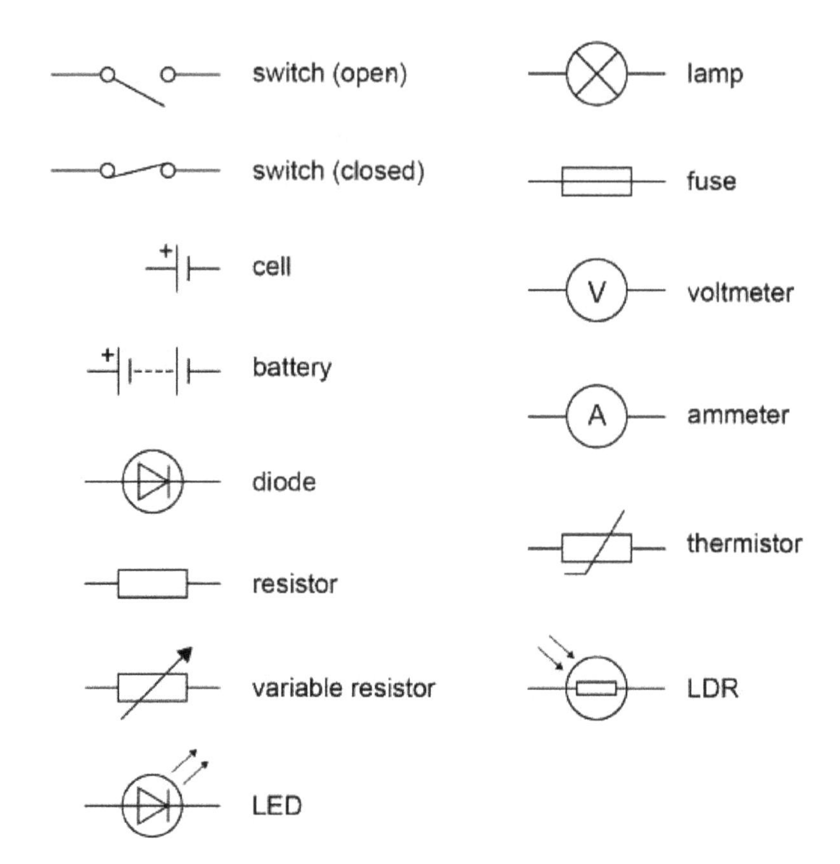

Appendix B: Simulation Software & CAD Software for Electronics

Computer-aided design (CAD) and simulation are important in modern engineering because they give a better idea of a problem and help increase efficiency. Using CAD software specifically designed for electrical engineering applications, engineers draw diagrams and schematics of electrical networks. Simulations also give a better idea of the circuit and instead of designing it physically, we can save our resources (money and time) with the help of simulations. Electrical simulation software is interactive and can help us find solutions to problems more easily. When we have an idea for a new application, the basic mathematical structure of that problem is the prerequisite. Just as engineers develop the mathematics of their models, developers use it in their models for interactive simulations. Below are a few helpful software tools that you are welcome to take a closer look at:

- **Multisim/Proteus:** For solutions of electronic and embedded circuits. Just like practical tools, these simulation software have tools like oscilloscope, function generator and multimeter for visualization.
- **Visio/Edraw and Max/AutoCAD electrical:** CAD software for electrical model representations.
- **Eagle/Altium Designer:** For professional PCB design
- **MATLAB/Mathematica:** For solving problems mathematically. MATLAB also has a built-in extension called Simulink, which is very useful in electrical simulations. Apart from simple mathematical solutions, MATLAB is a wide-ranging software that we can use to do almost anything: Signal analysis, power systems, electronics, energy management, engineering economics, and even robotics. Mathematica is also a comprehensive software and is particularly good for pure mathematical solutions.
- **ETAP:** For power system design, troubleshooting and load management.

For the circuits in this book the software **EveryCircuit (online) was** used. Feel free to google for it and try out this software!

Appendix C: A brief introduction to using an Arduino

Arduino is a commonly used microcontroller. Microcontrollers are complex programmable devices for control with simple instructions. With a microcontroller, we can control a simple system with a simple set of instructions. The Arduino compiler interprets Python and CPP programming languages into binary machine language and loads these instructions into the microcontroller. The Arduino is commonly referred to as a microcontroller, but it is not a microcontroller itself. It is a composite of various electronics with a microcontroller. Usually, the microcontroller used in Arduino is an 8-bit Atmel AT-mega. This makes Arduino an interactive and convenient microcontroller.

For more details on how to use an Arduino and step-by-step instructions, I recommend my book (expected to be published late 2021 / early 2022):

ARDUINO

Step by Step

The Ultimate Beginners' Guide

M.Eng. Johannes Wild

Closing words

Very good! You did it, you worked through the beginner course. Congratulations!

In this book I have tried to bring you the basic knowledge of electrical engineering and electronics simply explained closer. I hope that I have succeeded to some extent and that this book has brought you a well understandable and practical introduction to the world of electrical engineering!

The aim of this book was to bring you closer to how electrical engineering accompanies us in everyday life and what basic principles are involved. It should be a book that gives an understanding of electrical circuits and also an understanding of the most important components (e.g. resistor, transformer, capacitor, diode, etc.) in electrical engineering or electronics.

We have also covered the basics of DC and AC technology, their physical backgrounds and much more in this book!

With this basic course, you should now know everything you need to know as a beginner about the world of electrical engineering and electronics! Of course, it makes sense not to stop at this point and rather look into an advanced book to learn even more about the exciting subject of electrical engineering. However, if you can't do anything with this field (maybe you are more mechanically inclined) you have at least heard the basics now!

Together we have accomplished a lot in this course one way or another! Be rightly proud of yourself when you get to the end!

If you liked this book, I would be very happy if you leave me a rating and a short feedback, as well as recommend the book! Thank you very much!

Books you might also like

CPSIA information can be obtained
at www.ICGtesting.com
Printed in the USA
LVHW021450181122
733280LV00004B/229